Methods in Molecular Biology

Series Editor:
John M. Walker
School of Life and Medical Sciences
University of Hertfordshire
Hatfield, Hertfordshire, AL10 9AB, UK

For further volumes:
http://www.springer.com/series/7651

Mass Spectrometry of Proteins

Methods and Protocols

Edited by

Caroline A. Evans

Department of Chemical and Biological Engineering, University of Sheffield, Sheffield, South Yorkshire, UK

Phillip C. Wright

Faculty of Science, Agriculture & Engineering, Newcastle University, Newcastle upon lyne, UK

Josselin Noirel

Laboratoire GBCM (EA4627), Conservatoire National des Arts et Métiers HESAM Université, Paris, France

Editors
Caroline A. Evans
Department of Chemical and Biological Engineering
University of Sheffield
Sheffield, South Yorkshire, UK

Phillip C. Wright
Faculty of Science, Agriculture & Engineering
Newcastle University
Newcastle upon Tyne, UK

Josselin Noirel
Laboratoire GBCM (EA4627)
Conservatoire National des Arts et Métiers
HESAM Université
Paris, France

ISSN 1064-3745 ISSN 1940-6029 (electronic)
Methods in Molecular Biology
ISBN 978-1-4939-9234-8 ISBN 978-1-4939-9232-4 (eBook)
https://doi.org/10.1007/978-1-4939-9232-4

Library of Congress Control Number: 2019935881

This Humana Press imprint is published by the registered company Springer Science+Business Media, LLC, part of Springer Nature.
The registered company address is: 233 Spring Street, New York, NY 10013, U.S.A.

Preface

Technological advancement in the life sciences has fostered a paradigm shift in data generation that is collectively known as "omics." Systems biology, as it has come to be referred to, has become the primary perspective to adopt to make headway in precision medicine, biological engineering, etc. High-throughput data is systems biology's oxygen.

Systems biology represents a paradigm shift that requires the development of new protocols and new analytical tools in order to address fundamental questions that would have been otherwise out of reach. Like massively parallel sequencing (often referred to as next-generation sequencing, "NGS"), mass spectrometry (MS) is now at the center of countless applications allowing researchers to probe a very broad range of biological processes. As it happens, this is not the only common point with NGS:

- Many protocols exist, each one with a specific application/question at the end.
- Certain MS equipment will be more suited to certain applications (just like a Pacific Biosciences platform will be more suitable than an Illumina one, if longer reads are required)—even at a meta-level between protein analysis (proteomics) and metabolite analysis (metabolomics).
- Finally, in spite of a generally converging technology, data consolidation and analysis pipelines are still very much application dependent and likely to continue to evolve substantially in the coming years.

For all these reasons, it seemed important to offer an overview—a necessarily incomplete one, it must be said. Due to the importance of proteins in carrying out key cellular operations at the "business end" of biology, we have restricted our overview to mass spectrometry and proteomics. In particular, we focus on new pipelines and new workflows/protocols that allow for new frontiers in proteomics to be pushed outward facilitated by mass spectrometry.

New Biological Insights from Technological Breakthroughs

New opportunities arise with the emergence of better equipment: more precise, accurate, and faster mass spectrometers are a key factor driving the development of new methods. This has been crucial to the development of sequential windowed acquisition of all theoretical mass spectra, commonly known as SWATH-MS. It is one of the most promising approaches for proteomic mass spectrometry: improved accuracy and shortened duty cycle have allowed for this new analytical paradigm to achieve a broader dynamic range and to increase the number of peptides/proteins researchers are able to identify and to quantify. Lin et al. illustrate the use of SWATH-MS as an effective tool to detect biomarkers (Chapter 1).

Another aspect of proteomics that has substantially benefited from improved sensitivity and accuracy for mass spectrometry is the study of posttranslational modifications. Posttranslational modifications (PTMs) are key regulators of protein function. These often require specific enrichment strategies for detection and quantification due to the low stoichiometry of the modification.

Ubiquitination (Ub) is an important posttranslational process involving attachment of the ubiquitin molecule to lysine either on a substrate protein or on another ubiquitin molecule, leading to the formation of protein mono-, multi-, or polyubiquitination. The relative quantification of ubiquitination by using an mTRAQ-based chemical labelling strategy for the selective and enhanced detection of Ub-derived isopeptides is described (Chapter 2), applying SWATH-MS for data-independent acquisition and directed data mining. The generation of poly-Ub chains linked to a particular Ub lysine results in a chain of specific structural topology involved in the targeting of substrate proteins to different fates. Longworth and Dittmar describe a polyubiquitination (Chapter 3).

In terms of other PTM analyses, methods for enrichment of phosphopeptides, acetylated peptides, and palmitoylated peptides are described in stepwise detail in three chapters. Indeed, Barrios-Llerena and Bihan describe the use of custom titanium dioxide microcolumns for phosphopeptide enrichment coupled to label-free quantification for relative quantification (Chapter 4). Lund et al. detail an antibody-based enrichment of acetylated peptides (Chapter 5), and Woodley and Collins give an illustration of an acyl-biotin exchange (ssABE) protocol for analysis of S-acylation (Chapter 6). Combining this approach with a metabolic-labelling SILAC-based relative quantification workflow allows for both the large-scale identification of S-palmitoylation sites and quantitative profiling of *S*-palmitoylation site changes. The use of isotope labelling has enabled to assignment and quantification of newly synthesized acetyl groups to determine dynamically regulated and steady-state modifications, as outlined for acetylation by Lund et al. (Chapter 5). In general, while there is an array of methods directed toward specific biological questions, there are common sample processing steps. One of these, reduction and alkylation of cysteine residues, has been the subject of debate for its use on gel and gel-free proteomic analysis. The practical consequences and technical design considerations are discussed by Evans (Chapter 7).

There is increasing interest in the detection and characterization of covalent chemical adducts on proteins such as those formed by drug interactions. A general method for deciphering the chemical composition and site assignment of chemical adducts is described by Antinori et al. through the use of innovative mass spectrometry and bioinformatics approaches, exemplified by analysis of busulfan drug adducts (Chapter 8).

Besides their posttranslational modifications, many proteins function in complexes. The identification of the composition of these complexes offers insights into the mechanism behind their function. Federspiel and Cristea outline protocols for the optimization of lysis conditions, antibody evaluation, affinity purification, and, ultimately, identification of protein complexes from endogenous immunoaffinity purifications using quantitative mass spectrometry (Chapter 9). Such an approach overs advantages over the use of tagged and overexpressed bait proteins, which can generate artifacts arising from the presence of a tag on a protein and the higher abundance of the protein of interest (bait).

Increased sample throughput using mass spectrometers has given rise to completely new fields and utterly unexplored questions/biological processes. Just like intensive sequencing capabilities have led to the emergence of metagenomics and the prospect of studying non-model organisms almost in the wild, metaproteomics can reveal the intricacies of the bioprocesses taking place in ecosystems. Russo and Pandhal give a detailed account of how to study freshwater communities (Chapter 10) as an exemplar.

Dealing with Proteomics Data in a Big Data Era

New opportunities will also emerge from new paradigms in how downstream post-mass spectrometer analysis is performed. More data and more accurate data will call for new data pipelines and new ways of storing, exploiting, and integrating data, including data not obtained in one's own laboratory. This obviously includes new mathematical models that are developed to better process data but also the practices around data analysis. Fischer et al., for instance, present a pragmatic approach to peptide-to-protein summarization that attempts to identify and leverage the most trustworthy peptide quantifications in a proteomics dataset to derive protein quantifications. The rationale of the method is extrapolated from something proteomics practitioners are well aware of: low-intensity precursors carry much noise. But there are features other than just precursor intensity that are associated with low accuracy (e.g., charge). This shows that datasets are in a sense much richer than one could have imagined at first, and more complex analytical pipelines can be developed to harness that information for maximum benefit (Chapter 11).

In the current context where reproducibility (or lack thereof) in science has been recognized as a problem the scientific community needs to address, it has also become apparent that more complex analyses call for more transparency and more flexible ways to share and reuse both data and computational workflows. There's now a realization that more emphasis must be put on validation and robust experimental designs.

Reproducibility primarily depends on the robustness of the experimental design; Burzykowski et al. provide useful guidelines (Chapter 12). Central to that question is the number of replicates. Although sufficient biological replicates should be preferred in general, technical replicates can allow for the estimation of important noise parameters in advanced statistical models. At any rate, when considering a proteomics experiment, (non-metabolic) label-based approaches offer the most straightforward route to replication. Up to 8 or 11 samples can be simultaneously analyzed using iTRAQ or TMT labels, respectively. Pascovici et al. describe a pragmatic approach to increasing the number of samples so that both statistical power and lower false discovery rates can be achieved (Chapter 13).

Moreover, the ever-increasing complexity of individual studies cannot be denied: it is now common for a proteomics paper to make use of many complex data processing tools, libraries, and packages. Those pipelines are often tweaked too, making the most of the versatility and pervasiveness of languages such as R, Python, or Julia, to name a few.

There's an almost immediate upside to the (legitimate but perhaps daunting) demand for more transparent ways to share both the data and the method. The chapter written by Jarnuczak and colleagues provides a stimulating overview of the opportunities that arise as a consequence of the increasing amount of data made available (Chapter 14). Many research projects could benefit from public databases without the need of new experiments to be carried out, as is perhaps more common in other omics fields, such as genomics, transcriptomics, or epigenomics.

In terms of how the methods can be effectively communicated either to the purpose of proposing new computational methods or to the purpose of allowing the replication of a study, numerous solutions have been proposed. Two massively popular ones are covered in this book. First, Galaxy allows the researcher to create, adapt, and manage bioinformatics workflows for genomics, transcriptomics, and proteomics analyses. It uses a graphical interface in a web browser. Malmström illustrates some of the basic tasks that can be handled by Python in a chapter that is reminiscent of the "R for proteomics" documentation (Chapter 15). Stewart et al. illustrate the use of Galaxy for the identification of protein-protein interactions (Chapter 16). Second, Jupyter is a computational environment that lives too in the web browser through special files called "notebooks." Those notebooks are the cornerstone of many reproducible workflows in computational biology and beyond. Jupyter makes it possible to interact in many expressive ways with various "kernels," which are in charge of the computations and of the exploitation of the calculations. A popular kernel is powered by the programming language Python, which, through the efforts of a dynamic community, has expanded at a staggering pace. Many packages and libraries have been developed that make Python the go-to solution for advanced computational analysis.

Proteomics has come a long way in the last 10 or 15 years. It used to take weeks at the start of that timeframe to identify and perhaps quantify a few hundred proteins. Prior to even that, it had taken considerably longer to typically identify and quantify (usually by MALDI-TOF and 1D and 2D gels) 10s to low hundreds of proteins. Now, we are at a point that it merely takes hours to identify, quantify, and assess various posttranslational modifications at the rate of thousands. The majority of this usually happens end to end of a mass spectrometer and hyphenated upstream separation equipment, online. The trick is to maximize the robustness of this data and to maximize the information we obtain and not just the volume. This book, we believe, has gone some way in assessing the state of the frontier we are at and looking forward over this frontier to describe and seek new lands to explore.

Sheffield, South Yorkshire, UK *Caroline A. Evans*
Newcastle upon Tyne, UK *Phillip C. Wright*
Paris, France *Josselin Noirel*

Contents

Contributors

ADELINA E. ACOSTA-MARTIN • *biOMICS Biological Mass Spectrometry Facility, Faculty of Science Mass Spectrometry Centre, University of Sheffield, Sheffield, UK*
PAOLA ANTINORI • *Department of Clinical Neurosciences, Faculty of Medicine, University of Geneva, Geneva, Switzerland*
MARTIN E. BARRIOS-LLERENA • *Proteomics and Mass Spectrometry, Bioscience Core Labs, King Abdullah University of Science and Technology, Thuwal, Saudi Arabia*
ANDREW P. BECKERMAN • *Department of Animal and Plant Sciences, The University of Sheffield, Sheffield, UK*
THIERRY LE BIHAN • *Synthetic and Systems Biology, School of Biological Sciences, University of Edinburgh, Edinburgh, UK*
TOMASZ BURZYKOWSKI • *I-BioStat, Hasselt University, Diepenbeek, Belgium*
NAVIN CHICOOREE • *MS-Insight Ltd, Manchester, UK*
MAXEY C. M. CHUNG • *Department of Biochemistry, Yong Loo Lin School of Medicine, National University of Singapore, Singapore, Singapore*
JÜRGEN CLAESEN • *I-BioStat, Hasselt University, Diepenbeek, Belgium*
MARK O. COLLINS • *Department of Biomedical Science, Centre for Membrane Interactions and Dynamics (CMIAD), University of Sheffield, Sheffield, UK; Faculty of Science Mass Spectrometry Centre, University of Sheffield, Sheffield, UK*
NARCISO COUTO • *Department of Chemical and Biological Engineering, The University of Sheffield, Sheffield, UK*
ILEANA M. CRISTEA • *Department of Molecular Biology, Princeton University, Princeton, NJ, USA*
GUNNAR DITTMAR • *Proteomics of Cellular Signalling, Luxembourg Institute of Health, Strassen, Luxembourg*
CAROLINE A. EVANS • *Department of Chemical and Biological Engineering, University of Sheffield, Sheffield, South Yorkshire, UK*
JOEL D. FEDERSPIEL • *Department of Molecular Biology, Princeton University, Princeton, NJ, USA*
MARTINA FISCHER • *Bioinformatics Unit (MF1), Department for Methods Development and Research Infrastructure, Robert Koch Institute, Berlin, Germany*
BENJAMIN A. GARCIA • *Department of Biochemistry and Biophysics, Perelman School of Medicine, Epigenetics Institute, University of Pennsylvania, Philadelphia, PA, USA*
TIMOTHY J. GRIFFIN • *Department of Biochemistry, Molecular Biology and Biophysics, University of Minnesota, Minneapolis, MN, USA*
JOHN R. GRIFFITHS • *MS-Insight Ltd, Manchester, UK*
ERIC B. HAURA • *Department of Thoracic Oncology, Moffitt Cancer Center, Tampa, FL, USA*
PRATIK JAGTAP • *Department of Biochemistry, Molecular Biology and Biophysics, University of Minnesota, Minneapolis, MN, USA*

ANDREW F. JARNUCZAK • *European Molecular Biology Laboratory, European Bioinformatics Institute (EMBL-EBI), Cambridge, UK*
JAMES E. JOHNSON • *Minnesota Supercomputing Institute, University of Minnesota, Minneapolis, MN, USA*
YEKATERINA KORI • *Department of Biochemistry and Biophysics, Perelman School of Medicine, Epigenetics Institute, University of Pennsylvania, Philadelphia, PA, USA*
BRENT M. KUENZI • *Department of Drug Discovery, Moffitt Cancer Center, Tampa, FL, USA; Cancer Biology Ph.D. Program, University of South Florida, Tampa, FL, USA*
PRAVEEN KUMAR • *Department of Biochemistry, Molecular Biology and Biophysics, University of Minnesota, Minneapolis, MN, USA; Bioinformatics and Computational Biology, University of Minnesota, Minneapolis, MN, USA*
PIERRE LESCUYER • *Department of Genetic and Laboratory Medicine, Geneva University Hospitals, Geneva, Switzerland*
QIFENG LIN • *Department of Biochemistry, Yong Loo Lin School of Medicine, National University of Singapore, Singapore, Singapore*
JOSEPH LONGWORTH • *Proteomics of Cellular Signalling, Luxembourg Institute of Health, Strassen, Luxembourg*
PEDER J. LUND • *Department of Biochemistry and Biophysics, Perelman School of Medicine, Epigenetics Institute, University of Pennsylvania, Philadelphia, PA, USA*
LARS MALMSTRÖM • *Institute for Computational Science, University of Zurich, Zurich, Switzerland; S3IT, University of Zurich, Zurich, Switzerland; Division of Infection Medicine, Department of Clinical Sciences, Lund University, Lund, Sweden*
SUBINA MEHTA • *Department of Biochemistry, Molecular Biology and Biophysics, University of Minnesota, Minneapolis, MN, USA*
THÉO MICHELOT • *School of Mathematics and Statistics, University of Sheffield, Sheffield, UK*
MARK MOLLOY • *Australian Proteome Analysis Facility, Macquarie University, Sydney, NSW, Australia*
MARKUS MÜLLER • *SIB-Swiss Institute of Bioinformatics, University of Lausanne, Lausanne, Switzerland*
THILO MUTH • *Bioinformatics Unit (MF1), Department for Methods Development and Research Infrastructure, Robert Koch Institute, Berlin, Germany*
JAGROOP PANDHAL • *Department of Chemical and Biological Engineering, The University of Sheffield, Sheffield, UK*
DANA PASCOVICI • *Australian Proteome Analysis Facility, Macquarie University, Sydney, NSW, Australia*
BERNHARD Y. RENARD • *Bioinformatics Unit (MF1), Department for Methods Development and Research Infrastructure, Robert Koch Institute, Berlin, Germany*
DAVID A. RUSSO • *Department of Plant and Environmental Sciences, University of Copenhagen, Copenhagen, Denmark*
SIMONE SIDOLI • *Department of Biochemistry and Biophysics, Perelman School of Medicine, Epigenetics Institute, University of Pennsylvania, Philadelphia, PA, USA*
XIAOMIN SONG • *Australian Proteome Analysis Facility, Macquarie University, Sydney, NSW, Australia*
PAUL A. STEWART • *Department of Thoracic Oncology, Moffitt Cancer Center, Tampa, FL, USA; Biostatistics and Bioinformatics Shared Resource, Moffitt Cancer Center, Tampa, FL, USA*

HWEE TONG TAN • *Department of Biochemistry, Yong Loo Lin School of Medicine, National University of Singapore, Singapore, Singapore*
TOBIAS TERNENT • *European Molecular Biology Laboratory, European Bioinformatics Institute (EMBL-EBI), Cambridge, UK*
DIRK VALKENBORG • *I-BioStat, Hasselt University, Diepenbeek, Belgium*
JUAN ANTONIO VIZCAÍNO • *European Molecular Biology Laboratory, European Bioinformatics Institute (EMBL-EBI), Cambridge, UK*
KEITH T. WOODLEY • *Department of Biomedical Science, Centre for Membrane Interactions and Dynamics (CMIAD), University of Sheffield, Sheffield, UK*
JEMMA WU • *Australian Proteome Analysis Facility, Macquarie University, Sydney, NSW, Australia*
ZUO-FEI YUAN • *Department of Biochemistry and Biophysics, Perelman School of Medicine, Epigenetics Institute, University of Pennsylvania, Philadelphia, PA, USA*
THIRI ZAW • *Australian Proteome Analysis Facility, Macquarie University, Sydney, NSW, Australia*
XIAOLU ZHAO • *Hubei Key Laboratory of Cell Homeostasis, College of Life Sciences, Wuhan University, Wuhan, China*

Part I

New Biological Insights from Technological Breakthroughs

Chapter 1

Next Generation Proteomics for Clinical Biomarker Detection Using SWATH-MS

Qifeng Lin, Hwee Tong Tan, and Maxey C. M. Chung

Abstract

The technology of "sequential windowed acquisition of all theoretical fragment ion spectra," known as SWATH-MS, is rapidly gaining popularity as a next generation proteomics technology for comprehensive proteome quantitation. In this chapter, we describe the use of SWATH-MS as a label-free quantitative technique in a proteomics study to identify novel serological biomarker for colorectal cancer. We compared the secreted glycoprotein profiles (glyco-secretomes) enriched from the colon adenocarcinoma cell line HCT-116 and its metastatic derivative, E1, and observed that laminin β-1 (LAMB1) was oversecreted in E1 cells. This novel oversecretion of LAMB1 was validated in colorectal cancer patient serum samples, and ROC analyses showed that LAMB1 performed better than carcinoembryonic antigen (CEA) as a clinical diagnostic biomarker for colorectal cancer. We focus here on the sample preparation methodology and data processing workflow for SWATH-MS studies.

Key words Data-independent acquisition, SWATH-MS, Colorectal cancer, Secretome, Biomarker

1 Introduction

SWATH (sequential windowed acquisition of all theoretical fragment ion spectra)-MS [1] is a label-free quantitative technique that is rapidly gaining popularity. It has been purported as the next generation proteomics technology, and merges the high-throughput protein identification capabilities of discovery proteomics with the reproducible and accurate quantitation of targeted proteomics. This is achieved by using data-independent acquisition (DIA), which involves repeated acquisition of fragment ion spectra from all precursor ions within sequential series of 25 Da isolation windows throughout a specified mass range of 350–1250 m/z (termed as "swaths") [1]. The resulting data are composite spectra of the fragment ions derived from all the precursor ions within each swath. The peptides are identified from a spectra ion library

Caroline A. Evans et al. (eds.), *Mass Spectrometry of Proteins: Methods and Protocols*, Methods in Molecular Biology, vol. 1977, https://doi.org/10.1007/978-1-4939-9232-4_1,

that contains peptide identities with their corresponding fragment ion information, which is usually generated by traditional data-dependent acquisition (DDA; also known as information-dependent acquisition (IDA) on SCIEX instruments) and database searching. Quantitation is performed by area under the curve (AUC) measurement of the highest intensity fragment ion signals. Due to the high resolution capabilities of the quadrupole time-of-flight (QqTOF) instrument, the accuracy and reproducibility of SWATH-MS quantitative analysis is comparable to that of selected reaction monitoring (SRM) [2]. SWATH-MS is thus a powerful technology for unbiased quantitative proteomics in clinical biomarker detection. In this chapter, we illustrate the application of SWATH-MS to identify clinically relevant biomarkers for colorectal cancer diagnosis [3].

Colorectal cancer has the second highest incidence in the world, and accounts for 8% of all cancer-related deaths [4, 5]. Mortality in colorectal cancer is often due to metastasis, and liver metastasis is the most common form in the patients. The only clinical biomarker for monitoring of colorectal cancer recurrence and metastasis is the serum carcinoembryonic antigen (CEA). However, CEA levels are also affected in inflammatory conditions such as hepatitis and inflammatory bowel disease [6], and the prognostic utility of CEA for colorectal cancer remains controversial [7–10]. In order to identify novel colorectal cancer clinical biomarkers, we collected the conditioned media (CM) from human adenocarcinoma cell line HCT-116 and its metastatic derivative, E1, using the hollow fiber culture system (HFC). The HFC system has been previously reported as an efficient system for simultaneous collection and concentration of secreted proteins, coupled with reduced intracellular protein contamination [11, 12]. Furthermore, in an effort to delve deeper into the colorectal cancer secretome, we applied the multi-lectin affinity chromatography (MLAC) method [13] to enrich for glycoproteins in the HCT-116 and E1 secretomes (glyco-secretomes), which were then compared using SWATH-MS (workflow shown in Fig. 1). A total of 149 glycoproteins were found to be differentially secreted in E1 cells (Fig. 2), of which the novel oversecretion of laminin-β1 (LAMB1) was of particular note. We further showed using ELISA that LAMB1 levels were significantly higher in colorectal cancer patient sera as compared to healthy controls, and LAMB1 performed better than CEA in distinguishing between patients from controls (Fig. 3). As this chapter is focused on the SWATH-MS sample preparation and data processing workflow, the reader is recommended to refer to our previous publication [3] for more details on the procedure for CM collection and MLAC enrichment.

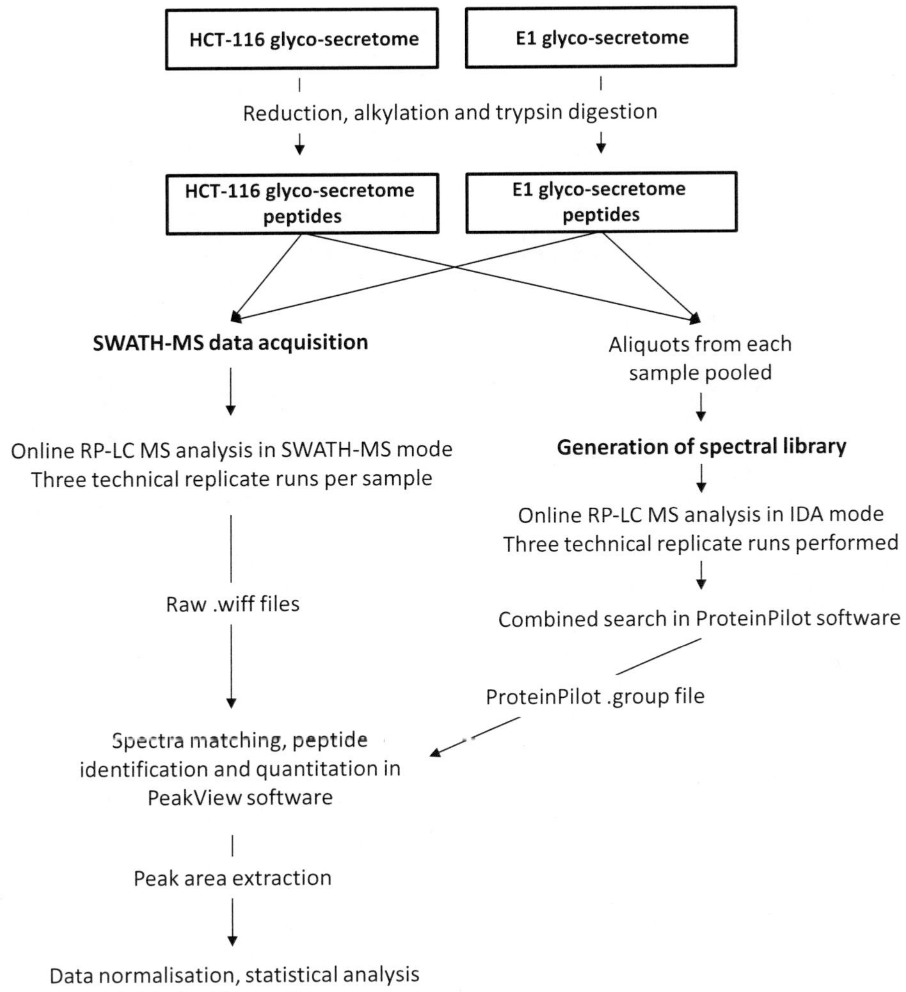

Fig. 1 SWATH-MS workflow outlining sample processing, data acquisition, and analysis procedures for identifying clinical colorectal cancer biomarkers. Glyco-secretomes are obtained by MLAC enrichment on CM collected from HCT-116 and E1 cells culture in the HFC system. After trypsin digestion, an aliquot is obtained from each sample, pooled, and then analyzed in data-dependent acquisition (DDA; known as IDA on SCIEX instruments) mode to generate a reference spectral library. The remaining samples are analyzed in SWATH-MS mode separately. Subsequently, peptide identification and quantitation is performed in the PeakView software, and the extracted peak areas are exported to the MarkerView software for normalization and statistical analysis

2 Materials

2.1 In-Solution Trypsin Digestion

1. 5 mM Tris-(2-carboxyethyl) phosphine (TCEP): dissolve Tris-(2-carboxyethyl) phosphine hydrochloride (Sigma-Aldrich) in ultrapure water.
2. 10 mM Methyl methane-thiosulfonate (MMTS): dissolve *S*-methyl methanethiosulfonate (Sigma-Aldrich) in isopropanol.
3. Sequencing grade modified trypsin.

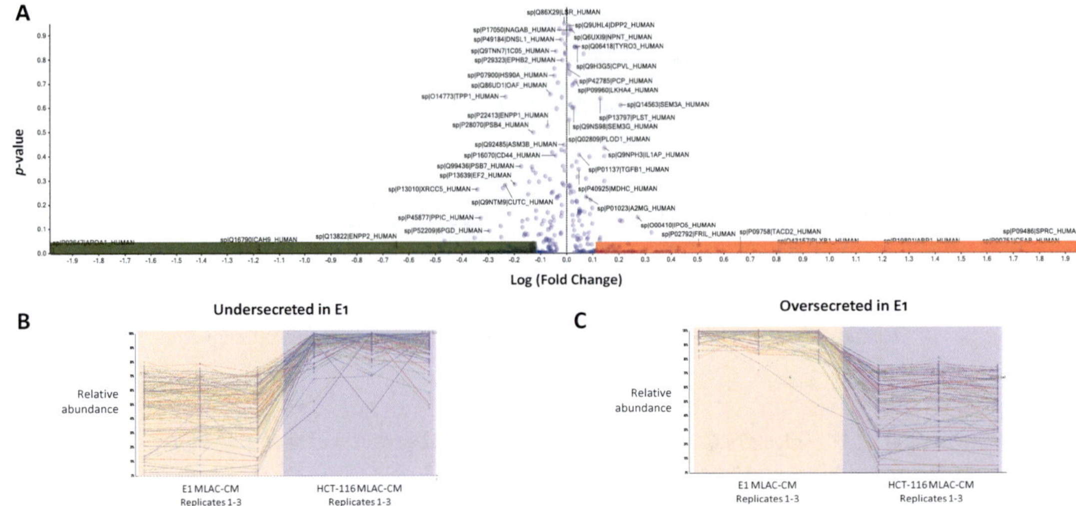

Fig. 2 SWATH-MS analysis of HCT-116 and E1 glyco-secretomes. (**a**) Visualization of the 421 quantified proteins in the HCT-116 and E1 glyco-secretomes in the form of a volcano plot. The logarithm of the fold-change ratio for each protein is plotted against its respective *p*-value. Using the cutoff criteria of fold-change >1.3 and *p*-value <0.05, 75 glycoproteins are observed to be undersecreted in E1 cells, as highlighted in green with their relative abundances depicted in (**b**). On the other hand, the oversecretion of 74 proteins is observed, as highlighted in red and their relative abundances plotted in (**c**)

4. Sep-Pak tC18 μElution Plate (Waters, Milford, MA).
5. Vacuum manifold (Millipore, Billerica, MA) with vacuum pump.
6. Acetonitrile (ACN), LC-MS grade.
7. 2% (v/v) ACN with 0.1% (v/v) formic acid (FA), LC-MS grade.
8. 80% (v/v) ACN with 0.1% (v/v) FA, LC-MS grade.

2.2 LC-MS Data Acquisition and Processing for SWATH-MS

1. NanoLC-Ultra system with cHiPLC-Nanoflex (Eksigent, Dublin, CA).
2. Precolumn, 200 μm × 0.5 mm, Reprosil-Pur C18-Aq, 3 μm 120 Å (Eksigent).
3. Analytical column, 75 μm × 15 cm, Reprosil-Pur C18-Aq, 3 μm 120 Å (Eksigent).
4. 2% (v/v) ACN with 0.1% (v/v) FA and 98% (v/v) ACN with 0.1% (v/v) FA.
5. TripleTOF 5600 Analyzer (AB SCIEX, Framingham, MA).
6. ProteinPilot software version 4.5 (AB SCIEX).
7. SwissProt *Homo sapiens* database (20,254 entries, 06 March 2013 release).
8. PeakView software version 1.2.03 with SWATH Acquisition MicroApp version 1.0.0.653 (AB SCIEX).
9. MarkerView software 1.2.1 (AB SCIEX).

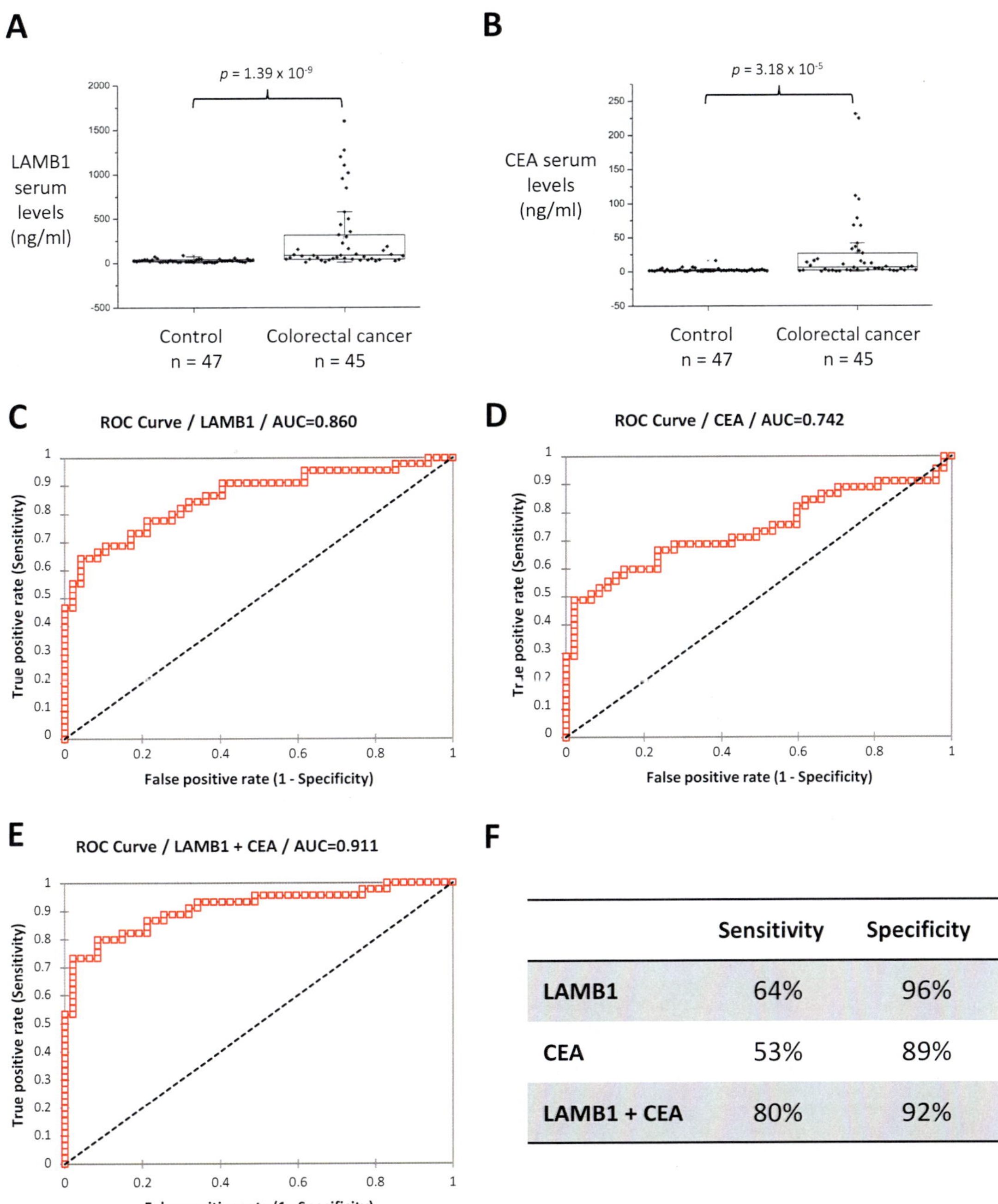

	Sensitivity	Specificity
LAMB1	64%	96%
CEA	53%	89%
LAMB1 + CEA	80%	92%

Fig. 3 LAMB1 levels are higher in colorectal cancer patient sera, and can be used for diagnosis. Levels of (**a**) LAMB1 and (**b**) CEA are measured in the serum samples from 45 colorectal cancer patient and 47 healthy individuals. LAMB1 levels are shown to be significantly higher in patient sera as compared to healthy controls. Box plot data are represented with the box denoting the upper and lower quartiles and interquartile range, the horizontal line showing the median, the dashes depicting middle 90% distribution, and the crosses representing the minimum and maximum values. The receiver operating characteristic (ROC) curve analysis is performed for LAMB1 (**c**) against CEA (**d**), and in combination (**e**), and the diagnostic performance (**f**) is compared using a cutoff of 58.0 ng/mL for LAMB1 and 5.0 ng/mL for CEA. Copyright (2015) Wiley. Used with permission (Lin et al., Analysis of colorectal cancer glyco-secretome identifies laminin beta-1 (LAMB1) as a potential serological biomarker for colorectal cancer, *Proteomics*, WILEY-VCH Verlag GmbH & Co)

2.3 ELISA Analysis on Clinical Patient Serum Samples

1. ELISA kit for LAMB1, SEA184Hu (USCN Life Sciences, Wuhan, China).
2. ELISA kit for CEA, BQ 062T (BQ Kits, San Diego, CA).
3. Phosphate-buffered saline (PBS): 137 mM NaCl, 2.7 mM KCl, 10 mM phosphate buffer.
4. Microplate sealing tape.
5. Thermomixer with microplate adaptor (Eppendorf, Hamburg, Germany).
6. ELx50 Microplate strip washer (BioTek, Winooski, VT).
7. Tecan Infinite M200 spectrophotometer (Tecan, Mannedorf, Switzerland).
8. OriginPro 8.6 software (OriginLabs, Northampton, MA).
9. XLSTAT v2.01 software (Addinsoft SARL, Paris, France).

3 Methods

3.1 In-Solution Trypsin Digestion

1. 30 μg total protein is obtained from MLAC-enriched HCT-116 and E1 CM samples each and reduced with 5 mM TCEP at 60 °C for 1 h, followed by alkylation with 10 mM MMTS at room temperature for 10 min.
2. The lyophilized trypsin vials are first reconstituted using the supplied resuspension buffer to a concentration of 0.8 μg/μL. The trypsin solution is then added to each sample at a trypsin-to-protein ratio of 1:10 (w/w). The samples are incubated at 37 °C for 16 h.
3. After trypsin digestion, the samples must be desalted prior to MS. First, the samples are diluted to 400 μL using 2% ACN, 0.1% FA.
4. The Sep-Pak tC18 μElution plate wells are activated twice using 400 μL of ACN, at full vacuum (approximately 15 Hg).
5. Next, the wells are equilibrated twice using 400 μL of 2% ACN, 0.1% FA, at full vacuum.
6. The samples are then added into the wells and allowed to stand for 5 min before applying low vacuum (approximately 5 Hg) (*see* **Note 1**).
7. The wells are washed twice with 400 μL of 2% ACN, 0.1% FA using low vacuum.
8. An empty 96-well plate is set up below the Sep-Pak tC18 μElution plate for elution. The peptides in each well are eluted using 100 μL of 80% ACN, 0.1% FA, at low vacuum (*see* **Note 2**).
9. After elution, the peptides are lyophilized using a vacuum evaporator and reconstituted in 2% ACN, 0.1% FA for LC-MS analysis (*see* **Note 3**).

10. Equal aliquots from each sample are combined to create a pooled sample, which will be analyzed separately to generate the reference spectral library.

3.2 LC-MS Data Acquisition and Processing for SWATH-MS

1. Online reversed phase (RP) LC separation prior to MS analysis is performed on the NanoLC Ultra system coupled with the cHiPLC-Nanoflex system (Eksigent).
2. From each sample, 5 μg of peptides are injected and first trapped on the 200 μm × 0.5 mm precolumn (Reprosil-Pur C18-Aq 3 μm 120 Å). Loading is performed at a flow rate of 3 μL/min for 6 min.
3. Subsequently, RP separation is performed on the 75 μm × 15 cm analytical column (Reprosil-Pur C18-Aq 3 μm 120 Å) at a flow rate of 300 nL/min. The RP solvent A is 2% ACN, 0.1% FA and RP solvent B is 98% ACN, 0.1% FA (*see* **Note 4**). The peptides are eluted using a linear gradient of 12–30% solvent B over 90 min. The column outlet is directly coupled to the nano ESI source of the SCIEX 5600 TripleTOF, such that eluted peptides are injected immediately for MS analysis.

3.2.1 Generation of Spectral Library

1. The pooled sample is analyzed in the IDA mode to generate the spectral library (*see* **Note 5**).
2. Precursor mass selection range is set to 350–1250 *m/z*, with 250 ms accumulation time per spectrum.
3. For MS/MS analysis, a maximum of 20 precursors per cycle is selected with dynamic exclusion for 15 s. MS/MS analysis is performed in high sensitivity mode with 100 ms accumulation time allowed for each precursor ion across a mass range of 100–1800 *m/z*. Collision energy (CE) is determined by the software based on the precursor ion charge with a spread of 5 eV according to the formula $|CE| = slope \times (m/z) + intercept$, where the slope and the intercept depend on the ion's charge (Table 1).

Table 1
Slopes and intercepts as a function of the charge

Charge	Slope	Intercept
Unknown	0.05	5
1	0.05	9
2	0.049	−1
3	0.048	−2
4	0.05	−2
5	0.05	−2

4. Three technical replicates of the MS analysis on the pooled samples are performed. Generated raw .wiff files are combined and searched using the ProteinPilot software with the Paragon algorithm against the SwissProt *Homo sapiens* database. Search parameters included cysteine alkylation by MMTS, all biological modifications specified by the algorithm, and detected protein threshold at 0.05.
5. FDR analysis is enabled and automatically performed by the software against a decoy database comprised of reversed protein sequences from the SwissProt database.
6. Resulting ProteinPilot .group file will be used as a reference spectral library for peptide matching in the SWATH-MS analyses.

3.2.2 SWATH-MS Analysis

1. CM samples are analyzed individually in the SWATH-MS mode. Three technical replicates are performed for each sample.
2. Precursors are selected with an accumulation time of 250 ms and across a mass range of 350–1250 m/z, in a set of 36 windows of 25 Da widths and 1 Da overlap (*see* **Note 6**).
3. Fragment ion spectra in each window are recorded with an accumulation time of 80 ms across a mass range of 100–1800 m/z in high sensitivity mode. The collision energy is fixed at 35 eV with 15 eV spread. The total cycle time is 2.98 s.
4. Raw .wiff files are analyzed using the SWATH Acquisition microapp in the PeakView software. The ProteinPilot group file generated in the previous section is imported as a reference library for spectral matching. A maximum of five peptides with three transitions is selected for each protein with the following processing parameters: 25 ppm ion library tolerance, 4 min extracted ion chromatogram (XIC) extraction window width, 0.01 Da XIC width (*see* **Note 7**). Only peptides with at least 99% confidence are considered; shared peptides or those with modifications are excluded.
5. For reliable quantitation, only proteins with three or more peptides available for quantitation are manually selected for XIC peak area extraction.
6. Extracted XIC peak areas are imported into the MarkerView software. Global normalization is performed according to the sum of the total areas of all detected and quantified proteins in the samples (*see* **Note 8**).
7. To identify proteins which have significantly different levels in the HCT-116 and E1 secretome, two-sample comparisons are performed using the Student's t-test in MarkerView. The averaged area sums of all the transitions derived from all the peptides for each protein are compared across the triplicate technical runs from each sample.

3.3 ELISA Analysis on Clinical Patient Serum Samples

1. For the LAMB1 ELISA measurements, patient and control serum samples are subjected to a 2× dilution using PBS (*see* **Note 9**). For the CEA ELISA kit, the serum samples are used without dilution.

3.3.1 LAMB1 ELISA

1. Each vial of the lyophilized recombinant LAMB1 standard is reconstituted with 1 mL of the Standard Diluent to make a 10 ng/mL stock. Using this, six dilutions ranging from 0.156 to 5 ng/mL are further prepared using serial dilution (*see* **Note 10**). A blank standard (0 ng/mL) should also be prepared with the Standard Diluent alone.
2. The Assay Diluent A is prepared by diluting 6 mL of 2× Assay Diluent A concentrate with 6 mL of ultrapure water. Similarly, 6 mL of 2× Assay Diluent B concentrate is diluted with 6 mL of deionized water to make Assay Diluent B.
3. The Wash Solution is prepared by adding 20 mL of the provided 30× Wash Solution to 580 mL of ultrapure water.
4. 100 μL of blank, standard, and diluted sera each are added into individual microplate wells. Sufficient volumes of each sample should be prepared for duplicate readings. The microplate is covered using a microplate sealing tape and incubated at 37 °C for 2 h.
5. The contents of each well are aspirated using the plate washer without any washing.
6. Sufficient volume of Detection Reagent A (100 μL required per well) is prepared by adding the stock Detection Reagent A into Assay Diluent A at a dilution ratio of 1:100. 100 μL of Detection Reagent A is then added to each well. The microplate is covered using a microplate sealing tape and incubated at 37 °C for 1 h.
7. Using the plate washer, the contents of each well are first aspirated. 350 μL of Wash Solution is then added to each well, allowed to sit for 1 min, before aspiration again. Repeat two more times.
8. Sufficient volume of Detection Reagent B (100 μL required per well) is prepared by adding the stock Detection Reagent B into Assay Diluent B at a dilution ratio of 1:100. 100 μL of Detection Reagent B is then added to each well. The microplate is covered using a microplate sealing tape and incubated at 37 °C for 30 min.
9. Using the plate washer, the contents of each well are first aspirated. 350 μL of Wash Solution is then added to each well, allowed to sit for 1 min, before aspiration again. Repeat four more times.

10. 90 μL of TMB Substrate Solution is added to each well. The plate is covered using a microplate sealing tape and incubated at 37 °C for 15 min (*see* **Note 11**). 50 μL of Stop Solution is added to each well.
11. The bottom of the microplate is cleaned using a piece of tissue paper before measuring the absorbance at 450 nm using the microplate reader.
12. Duplicate readings for each standard and serum sample are averaged, followed by normalization against the averaged blank (0 ng/mL) reading. The standard curve is then constructed by plotting LAMB1 concentration against the normalized absorbance readings in MS Excel. A second-order polynomial regression line through the origin is fitted to the data points. This is subsequently used to calculate the LAMB1 concentration in the serum samples with the dilution factor taken into account (*see* **Note 12**).

3.3.2 CEA ELISA

1. The Wash Buffer is prepared by adding 25 mL of the 20× Wash Concentrate to 475 mL of ultrapure water.
2. The CEA standards ranging from 0 to 50 ng/mL are provided ready-to-use. 50 μL of standards and sera are added into individual microplate wells.
3. 100 μL of the Enzyme Conjugate is added to each well. The microplate is covered using a microplate sealing tape and incubated at room temperature for 1 h.
4. Using the plate washer, the contents of each well are removed and 300 μL of Wash Buffer is added. This is repeated three more times.
5. 100 μL of TMB Substrate is added to each well and allowed to incubate at room temperature for 10 min. Subsequently, 50 μL of Stop Solution is added to each well.
6. The bottom of the microplate is cleaned using a piece of tissue paper before measuring the absorbance at 450 nm using the microplate reader.
7. Duplicate readings for each standard and serum sample are averaged, followed by normalization against the averaged blank (0 ng/mL) reading. The standard curve is then constructed by plotting CEA concentration against the normalized absorbance readings in MS Excel. A second-order polynomial regression line through the origin is fitted to the data points. This is subsequently used to calculate the CEA concentration in the serum samples (*see* **Note 12**).

3.3.3 Data Visualization and Diagnostic Performance Analysis

1. Box plots to visualize LAMB1 and CEA levels in patient and control sera are constructed using the Box Chart function in the OriginPro software.
2. Individual ROC curves for LAMB1 and CEA levels are generated using the ROC Curves function in the XLSTAT add-on for MS Excel with default parameters. The recommended cut-off LAMB1 and CEA levels for optimal diagnostic sensitivity and specificity are determined by the software.
3. To determine the diagnostic performance of LAMB1 and CEA when used in combination, factor analysis of LAMB1 and CEA sera levels is first performed. The resultant factor variables are then used to generate the ROC curve and diagnostic parameters for the combined LAMB1 and CEA test.

4 Notes

1. The flowthrough in this step can be passed through the Sep-Plate again to ensure efficient binding of the peptides.
2. To ensure that all bound peptides are completely recovered, another round of elution can be performed with 100 μL of 80% ACN, 0.1% FA and combined with the first set.
3. Quantitation of the total peptide content in each sample can be performed after reconstitution. This will ensure that the subsequent injection amount for LC-MS analysis is equal across all samples. Estimation of the peptide content may be performed using absorbance at 214 nm or through colorimetric assays (*e.g.* Pierce Quantitative Colorimetric Peptide Assay).
4. All chemicals used for preparation of LC solvents should be LC-MS grade. LC solvents should be degassed prior to use.
5. Mass calibration should be performed before MS analysis to ensure that the mass accuracy is within 2 ppm. Throughout analysis of the samples, automatic mass calibration should be performed approximately every 6 h.
6. Instead of fixed 25 Da windows, variable windows which are adjusted based on the precursor mass density can also be used. Smaller width windows are used when at precursor mass dense *m/z* regions, while wider windows are used at *m/z* ranges with fewer precursors. This effectively improves the SWATH-MS coverage without compromising the quantitative reproducibility.
7. Although not used in this study, the current version of SWATH Acquisition microapp allows for chromatographic retention time alignment among the SWATH-MS samples. Since retention time is one of the parameters used for spectral matching,

alignment is highly recommended especially for long multiple sample runs with potential for retention time drift. This may be performed using spiked-in calibration standards which are commercially available (iRT kit from Biognosys, Switzerland). Alternatively, a range of endogenous peptides (around six to eight) which span the entire chromatographic time range could be manually selected as retention time calibration standards.

8. Global normalization using total area sums is based on the presumption that the total protein abundances in the compared samples are largely similar. However, there might be some instances where this might not be applicable (*e.g.* comparison between cell lysates and secretomes). In such circumstances, normalization may be performed using spiked-in internal standards such as synthetic peptides.
9. All handling of serum samples should be performed in a Class II Biological Safety Cabinet. Dilution of serum samples, if required, should be performed in separate tubes instead of directly in the ELISA plate wells.
10. Serial dilution preparation of standards should be performed in separate tubes instead of directly in the ELISA plate wells.
11. This incubation step should be performed in the dark. The ELISA plate should be monitored occasionally to ensure that the colorimetric reaction in the wells have not reached saturation. The total incubation time must not exceed 30 min.
12. If there are any serum sample readings that exceed the range of the standard curve, these samples must be further diluted before repeating the ELISA measurement.

Acknowledgments

Q.L. acknowledges the support of a National University of Singapore research scholarship. The authors also acknowledge the funding from the National Medical Research Council, Singapore (NMRC grant 1217/2009).

References

1. Gillet LC, Navarro P, Tate S et al (2012) Targeted data extraction of the MS/MS spectra generated by data-independent acquisition: a new concept for consistent and accurate proteome analysis. Mol Cell Proteomics 11(6):O111.016717
2. Liu Y, Huttenhain R, Surinova S et al (2013) Quantitative measurements of N-linked glycoproteins in human plasma by SWATH-MS. Proteomics 13(8):1247–1256
3. Lin Q, Lim HS, Lin HL et al (2015) Analysis of colorectal cancer glyco-secretome identifies laminin beta-1 (LAMB1) as a potential serological biomarker for colorectal cancer. Proteomics 15(22):3905–3920
4. Bray F, Ren JS, Masuyer E, Ferlay J (2013) Global estimates of cancer prevalence for 27 sites in the adult population in 2008. Int J Cancer 132(5):1133–1145

5. Jemal A, Bray F, Center MM et al (2011) Global cancer statistics. CA Cancer J Clin 61(2):69–90
6. van der Schouw YT, Verbeek AL, Wobbes T et al (1992) Comparison of four serum tumour markers in the diagnosis of colorectal carcinoma. Br J Cancer 66(1):148–154
7. Carpelan-Holmstrom M, Haglund C, Lundin J et al (1996) Pre-operative serum levels of CA 242 and CEA predict outcome in colorectal cancer. Eur J Cancer 32A(7):1156–1161
8. Carriquiry LA, Pineyro A (1999) Should carcinoembryonic antigen be used in the management of patients with colorectal cancer? Dis Colon Rectum 42(7):921–929
9. Harrison LE, Guillem JG, Paty P, Cohen AM (1997) Preoperative carcinoembryonic antigen predicts outcomes in node-negative colon cancer patients: a multivariate analysis of 572 patients. J Am Coll Surg 185(1):55–59
10. Moertel CG, O'Fallon JR, Go VL et al (1986) The preoperative carcinoembryonic antigen test in the diagnosis, staging, and prognosis of colorectal cancer. Cancer 58(3):603–610
11. Chiu KH, Chang YH, Liao PC (2013) Secretome analysis using a hollow fiber culture system for cancer biomarker discovery. Biochim Biophys Acta 1834(11):2285–2292
12. Wu HY, Chang YH, Chang YC, Liao PC (2009) Proteomics analysis of nasopharyngeal carcinoma cell secretome using a hollow fiber culture system and mass spectrometry. J Proteome Res 8(1):380–389
13. Yang Z, Hancock WS (2004) Approach to the comprehensive analysis of glycoproteins isolated from human serum using a multi-lectin affinity column. J Chromatogr A 1053(1–2): 79–88

Chapter 2

A Combined Chemical Derivatization/Mass Spectrometric Method for the Enhanced Detection and Relative Quantification of Protein Ubiquitination

Navin Chicooree and John R. Griffiths

Abstract

Mass spectrometry (MS) is a sensitive analytical technique with wide application across the sciences including for the detection of peptides and proteins in biological analysis. Ubiquitinated (Ub) proteins are typically analyzed by proteolytic digestion and subsequent chromatographic separation followed by MS detection of the resulting isopeptides. Here we describe a novel method which enables enhanced detection of this important posttranslational modification (PTM) by use of a simple chemical labeling strategy prior to Data-Independent Acquisition (DIA) using a SWATH-based acquisition approach on a suitable Quadrupole-Time-Of-Flight (Q-TOF) mass spectrometer.

Key words SWATH, Data-independent acquisition, Ubiquitin, Posttranslational modifications, Mass spectrometry, Diagnostic ions

1 Introduction

Protein ubiquitination is an important reversible posttranslational modification involved in many cellular processes [1–4]. Dysregulation of the ubiquitination pathway has been implicated in a number of human diseases such as the cancers [5–7]. Detection of ubiquitination using mass spectrometry has typically taken advantage of the diglycine (–GG) iso-chain remnant, which remains attached to the modified lysine residue upon digestion with trypsin, resulting in a mass addition of 114.0429 Da for these isopeptides [8]. Utilizing this mass addition as a variable modification in database search routines is, however, prone to false positive assignments for a number of reasons. Firstly, the mass difference of +114.0429 Da is identical to the addition of an asparagine amino acid residue. Secondly, potential interference

Caroline A. Evans et al. (eds.), *Mass Spectrometry of Proteins: Methods and Protocols*, Methods in Molecular Biology, vol. 1977,
https://doi.org/10.1007/978-1-4939-9232-4_2,

has been reported from the use of iodoacetamide for cysteine alkylation [9]. Thirdly, there is a lack of corroborating modification-specific ions released upon collision-induced dissociation (CID) of these isopeptides [10].

We previously reported [11] on the use of reductive methylation (dimethyl labeling) to enable improved detection of isopeptides derived from ubiquitinated proteins. Dimethyl labeling modifies primary amines such as lysine residues and the N-termini of peptides, rendering their CID-derived fragment ions more stable [12]. Dimethyl labeling also enables relative quantification to be achieved in the MS domain between different (up to triplex) samples [13]. Since the –GG iso-chain remnant imparts a second N-terminus to the isopeptides, diagnostic fragment ions were shown to be present in their MS/MS spectra after such derivatization [11].

Here, we present a significant improvement on previous strategies by using a combination of mTRAQ chemical derivatization with SWATH (DIA) acquisition [14] on a Q-TOF mass spectrometer for the selective and enhanced detection of Ub-derived isopeptides (Fig. 1). The approach, termed Mass spectral Enhanced Detection of Ubls using SWATH Acquisition (MEDUSA) [15, 16], is capable of enhancing the detection of isopeptides (Fig. 2) and of delivering relative quantification across three channels. Also, since the mTRAQ labels are non-isobaric, and labeling takes place at the N-termini of peptides, quantification is achievable in both the precursor and product ion spectral domains (Fig. 3). In principle, the method is applicable to high resolution, accurate mass TOF-based mass spectrometers. For readers with an interest in ubiquitination, an accompanying chapter by Longworth and Dittmar details a methodology for analysis of poly-ubiquitination chain topology by PRM.

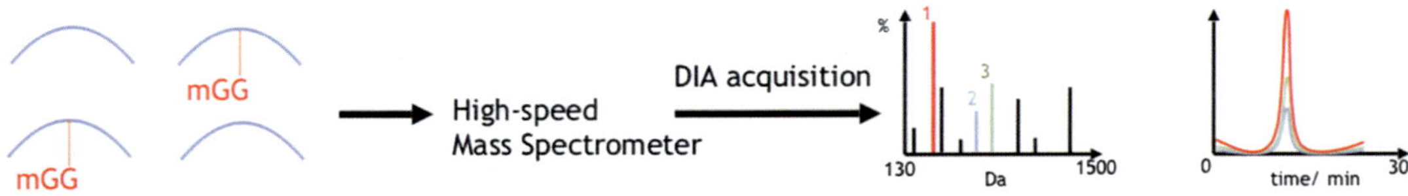

Fig. 1 Schematic representation of the MEDUSA workflow. Ub diagnostic ions are generated when mTRAQ-labeled isopeptides are subjected to collision-induced dissociation. SWATH DIA data are then interrogated for the presence of these ions, which are subsequently used for quantitation. Source: Griffiths et al. [15]. Reproduced with permission from Springer

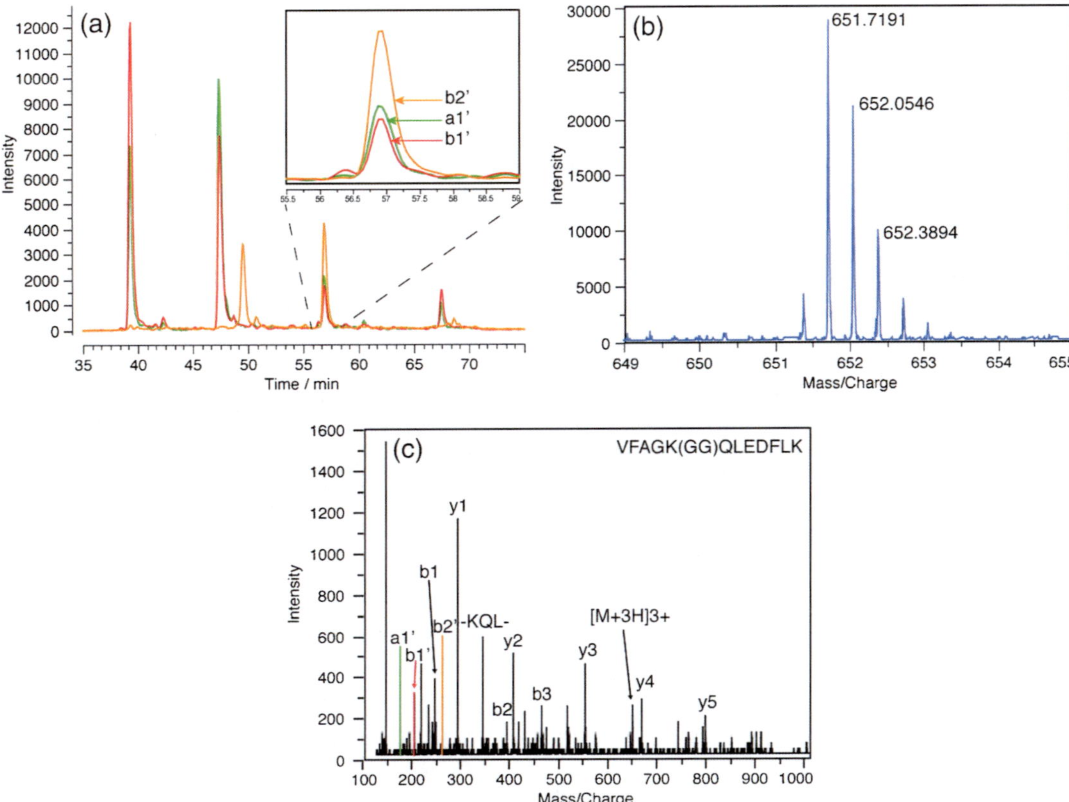

Fig. 2 (**a**) Overlay of extracted diagnostic ion (GG-tag, Δ8) chromatograms in SWATH window 649–655 Da. Inset shows perfect chromatographic overlay of the three GG-tag diagnostic ions corresponding to isopeptide; VFAGKGGQLEDFLK (+3); RT = 56.92 min; *m/z* 651.7176. (**b**) MS spectrum at 56.90 min in the range 649–655 Da. (**c**) MS/MS spectrum for SWATH window 649–655 Da at RT = 56.92 min. Source: Griffiths et al. [15]. Reproduced with permission from Springer

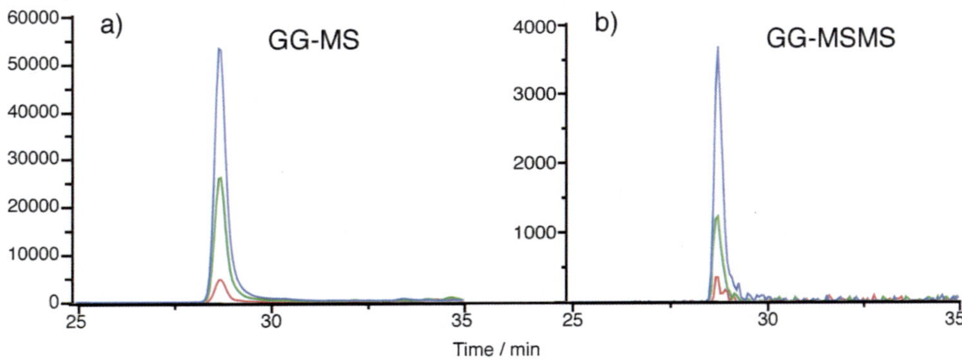

Fig. 3 (**a**) MS and (**b**) MS/MS-based quantitation of mTRAQ-labeled isopeptides based upon extracted ion peak areas. Lower overlaid curves correspond to Δ0-tagged (100 fmol), middle overlaid curves relate to Δ4-tagged (500 fmol), and upper overlaid curves represent Δ8-tagged (1 pmol) isopeptides. Source: Griffiths et al. [15]. Reproduced with modification with permission from Springer

2 Materials

All solutions should be freshly prepared using solvents and reagents of HPLC grade or better. It is essential to observe all safety precautions as outlined in the mTRAQ labeling kit [17].

2.1 Sample Preparation

1. SpeedVac evaporator.
2. Benchtop vortex mixer.
3. Benchtop centrifuge with pulse-spin function.
4. Eppendorf tubes.
5. Heating block or water bath.
6. Tetraethylammonium bicarbonate (TEAB): 500 mM solution in water as digestion buffer (*see* **Note 1**).
7. Sequencing grade modified porcine trypsin (lyophilized and stored at 4 °C) (*see* **Note 2**).
8. Tris (2-carboxyethyl)phosphine (TCEP)—disulfide bond reduction reagent from mTRAQ kit (*see* **Note 3**).
9. Methyl methanethiosulfonate (MMTS)—cysteine-blocking reagent from mTRAQ kit (*see* **Note 4**).
10. mTRAQ labeling reagents (*see* **Note 5**).
11. Trifluoroacetic acid (TFA): 0.1% solution in water (*see* **Note 6**).

2.2 LC-MS/MS

1. Mobile phase Buffer A: 0.1% formic acid in water. Add 1 mL formic acid to 999 mL water (*see* **Note 7**).
2. Mobile phase Buffer B: 0.1% formic acid in acetonitrile. Add 1 mL formic acid to 999 mL acetonitrile.
3. Mass spectrometer: 5600 or 6600 TripleTOF (Quadrupole-Time-of-Flight) mass spectrometer equipped with nanospray ion source interface.
4. Liquid chromatography: Nanoflow chromatography with C18 capillary column, autosampler, dual pumps, and column oven (*see* **Note 8**).

2.3 Data Processing Software

1. Peakview software ver. 1.1.0.1 or later (SCIEX).
2. Mascot search engine or equivalent (Matrix Science) [18].
3. Skyline [19].

3 Methods

The following procedure is suitable for the triplex quantitative comparison of 3 × 100 μg protein samples.

3.1 Protein Digestion and mTRAQ Labeling

1. To 100 μg of lyophilized protein sample in an Eppendorf tube, add 20 μL of 500 mM TEAB containing 4 μg of trypsin (*see* **Note 2**). Vortex briefly and leave to digest for 16 h (overnight) at 37 °C with shaking.
2. After overnight digestion, briefly pulse-spin the tube to ensure no sample remains in the cap, and add a further 4 μg of trypsin. Vortex briefly and leave to digest for a further 4 h at 37 °C with shaking.
3. Briefly pulse-spin sample and add 2 μL of 50 mM TCEP to reduce cysteine disulfide bonds. Incubate at 60 °C for 1 h.
4. Briefly pulse-spin sample and allow to cool to room temperature. Add 1 μL of 200 mM MMTS to block sulfhydryl groups of the cysteine residues. Incubate at room temperature for 10 min.
5. Pulse-spin sample.
6. **Steps 1–5** above should be repeated for all three samples to be compared.
7. Allow one vial each of Δ0, Δ4, and Δ8 mTRAQ-labeling reagent to reach room temperature and pulse-spin to ensure all contents are at the bottom of the tubes.
8. Add 50 μL of isopropanol to each tube, vortex briefly, then pulse-spin.
9. Transfer the contents of each tube to the three digested protein samples prepared above and vortex briefly to mix, then pulse-spin.
10. Incubate the tubes at room temperature for 1 h.
11. Dry samples on a SpeedVac evaporator and store at −20 °C until ready to analyze by LC-MS/MS.

3.2 LC-MS/MS Analysis and Data Acquisition

1. Reconstitute the mTRAQ-labeled peptide samples into 10 μL of 0.1% TFA.
2. Mix together the three samples of Δ0, Δ4, and Δ8 mTRAQ-labeled peptides, briefly vortex, and pulse-spin. Place into a 1 mL autosampler vial.
3. Using the chromatographic conditions in **Note 8**, or similar, inject 2 μL of the mixed labeled-peptide sample and acquire the MS data using a SWATH acquisition (*see* **Note 9**) on a suitable Q-TOF mass spectrometer (*see* **Note 10**).

3.3 Post-acquisition Data Analysis

3.3.1 Identification of Ub-Isopeptides

1. Diagnostic low mass product ions for Ub-derived isopeptides are generated upon CID in a Q-TOF mass spectrometer (Table 1). Enhanced detection of these isopeptides is achieved by manual data interrogation.
2. In order to achieve enhanced detection via comprehensive structural elucidation and identification of a putative isopep-

Table 1
Accurate masses of diagnostic ions generated from Δ0, Δ4, and Δ8 mTRAQ-labeled isopeptides

	GG-tag, Δ0 mTRAQ	GG-tag, Δ4 mTRAQ	GG-tag, Δ8 mTRAQ
a1′ ion	*m/z* 170.1288	*m/z* 174.1359	*m/z* 178.1430
b1′ ion	*m/z* 198.1237	*m/z* 202.1308	*m/z* 206.1379
b2′ ion	*m/z* 255.1452	*m/z* 259.1523	*m/z* 263.1594

tide, first, extract the three low mass diagnostic product ions individually from each MS/MS corresponding SWATH window using the Peakview software or Skyline (*see* **Note 11**).

3. Second, overlay these three extracted diagnostic product ion chromatograms with each other. Confirmation of an isopeptide is achieved when extracted a1′, b1′, and b2′ low mass diagnostic product ions overlay perfectly, thus demonstrating perfect chromatographic retention time co-elution (Fig. 2a).
4. Third, the precursor isopeptide ion representing the corresponding MS scan at the identical retention time of the three low mass diagnostic product ions enables the *m/z* and charge state of the precursor isopeptide ion to be obtained (Fig. 2b).
5. Since full MS/MS fragmentation is obtained along with the three low mass diagnostic product ions, enhanced detection of the isopeptide is rendered possible (Fig. 2c).

3.3.2 MS and MS/MS Quantitation of Ub-Isopeptides

1. In order to achieve MS-based quantitation, extract each of the three differentially labeled monoisotopic isopeptide precursor ion peaks (Δ0, Δ4, and Δ8) individually by following **steps 1–4** above (Fig. 3a). Integrate the peak areas for each of the differentially labeled isopeptide precursor isopeptide ion peaks Δ0, Δ4, and Δ8) using Peakview software or Skyline.
2. In order to achieve MS/MS-based quantitation, extract and then integrate the dominant b2′ ions generated upon CID (Fig. 3b) using Peakview software or Skyline.

4 Notes

1. Mix 500 μL of 1 M TEAB with 500 μL water using a benchtop vortex mixer for a 500 mM TEAB solution. Prepare fresh on day of use.
2. Trypsin should be dissolved in digestion buffer (500 mM TEAB) as required and used immediately. Add 100 μL

500 mM TEAB to a 20 μg tube of lyophilized trypsin, giving a trypsin concentration of 200 μgmL^{-1}.

3. 50 mM TCEP is supplied in mTRAQ kit as reducing reagent. May be prepared from alternative supplier and made to desired concentration in water. Observe all safety precautions contained in the mTRAQ kit documentation [17].
4. 200 mM MMTS is supplied in mTRAQ kit as cysteine-blocking reagent. May be prepared from alternative supplier and made to desired concentration in isopropanol. Observe all safety precautions contained in the mTRAQ kit documentation [17]. Prepare in a fume cupboard.
5. mTRAQ labeling reagent exists in three forms, Δ0, Δ4, and Δ8. The peptides are derivatized at lysine residues and N-termini, and each label adds a mass differing by the number of additional stable isotopes (C13, N15, and O18) contained in each reagent, with the Δ0 reagent containing only C12, N14, and O16 [17].
6. Trifluoroacetic acid must be handled in a fume cupboard wearing suitable gloves and safety goggles. Fumes are toxic and corrosive.
7. Formic acid must be handled in a fume cupboard wearing suitable gloves and safety goggles. Fumes are corrosive.
8. A suitable chromatographic gradient for this analysis would be a linear gradient from 5% to 30% Buffer B over a 60-min period at a flow rate of 300 nLmin^{-1} using C18 analytical column with dimensions of 250 mm × 0.075 mm × 3 μm particle size. Column may be maintained at 60 °C to reduce backpressure. Peptides should first be trapped onto a C18 cartridge such as 0.5 mm × 0.2 mm × 3 μm particle size, at a flow rate of 3 μLmin^{-1} for 20 min. Suitable loading buffer is 97.9% water +2% acetonitrile +0.1% formic acid.
9. A suitable SWATH cycle consists of one TOF MS full scan (100 ms) and 80 × 6 Da window size MS/MS acquisitions of 40 ms giving a total cycle time of 3.3 s.
10. Currently SWATH acquisition is enabled on SCIEX (Foster City, CA, USA) Q-TOF type mass spectrometers only. However, other TOF-based mass spectrometers may also be used for data acquisition and interrogation, software permitting. Furthermore, the labeling strategy may also be employed in data-dependent analyses using any commercial instrument with sufficient MS/MS resolution (minimum 20,000) and searching the data using a suitable search engine such as Mascot [18].
11. Skyline software [19] is a freely available quantitative software tool suitable for both DDA and DIA data acquisition.

Acknowledgments

We would like to acknowledge our coauthors on the original paper [15], for their equal contribution to the development of the MEDUSA workflow described herein.

References

1. Finley D, Chau V (1991) Ubiquitination. Annu Rev Cell Biol 7:25–69
2. Mukhopadhyay D, Riezman H (2007) Proteasome-independent functions of ubiquitin in endocytosis and signaling. Science 315:201–205
3. Bergink S, Jentsch S (2009) Principles of ubiquitin and SUMO modifications in DNA repair. Nature 458:461–467
4. Hochstrasser M (2009) Origin and function of ubiquitin-like proteins. Nature 458:422–429
5. Sun Y (2006) E3 ubiquitin ligases as cancer targets and biomarkers. Neoplasia 8:645–654
6. Bologna S, Ferrari S (2013) It takes two to tango: ubiquitin and SUMO in the DNA damage response. Front Genet 4:1–18
7. Chen Z, Lu W (2015) Roles of ubiquitination and SUMOylation on prostate cancer: mechanisms and clinical implications. Int J Mol Sci 16:4560–4580
8. Peng J, Schwartz D, Elias JE et al (2003) A proteomics approach to understanding protein ubiquitination. Nat Biotechnol 21(8): 921–926
9. Nielsen ML, Vermeulen M, Bonaldi T et al (2008) Iodoacetamide-induced artifact mimics ubiquitination in mass spectrometry. Nat Methods 5:459–460
10. Warren MRE, Parker CE, Mocanu V et al (2005) Electrospray ionization tandem mass spectrometry of model peptides reveals diagnostic product ions for protein ubiquitination. Rapid Commun Mass Spectrom 19:429–437
11. Chicooree N, Connolly Y, Tan C-T et al (2013) Enhanced detection of isopeptides using reductive methylation. J Am Soc Mass Spectrom 24:421–430
12. Fu Q, Li L (2006) Fragmentation of peptides with N-terminal dimethylation and imine/methylol adduction at the tryptophan side-chain. J Am Soc Mass Spectrom 17:859–866
13. Hsu JL, Huang SY, Chow NH et al (2003) Stable-isotope dimethyl labeling for quantitative proteomics. Anal Chem 75:6843–6852
14. Gillet LC, Navarro P, Tate S et al (2012) Targeted data extraction of the MS/MS spectra generated by data-independent acquisition: a new concept for consistent and accurate proteome analysis. Mol Cell Proteomics 11:O111.016717
15. Griffiths JR, Chicooree N, Connolly Y et al (2014) Mass spectral enhanced detection of Ubls using SWATH acquisition: MEDUSA—simultaneous quantification of SUMO and ubiquitin-derived isopeptides. J Am Soc Mass Spectrom 25:767–777
16. Chicooree N, Smith DL (2016) Mass spectrometry methods for the analysis of isopeptides generated from mammalian protein ubiquitination and SUMOylation. In: Griffiths JR, Unwin RD (eds) Analysis of protein post-translational modifications by mass spectrometry. John Wiley and Sons, Hoboken, NJ
17. https://sciex.com/Documents/Downloads/Literature/mass-spectrometry-4373841D.pdf
18. Perkins DN, Pappin DJ, Creasy DM et al (1999) Probability-based protein identification by searching sequence databases using mass spectrometry data. Electrophoresis 20: 3551–3567
19. MacLean B, Tomazela DM, Shulman N et al (2010) Skyline: an open source document editor for creating and analyzing targeted proteomics experiments. Bioinformatics 26: 966–968

Chapter 3

Assessment of Ubiquitin Chain Topology by Targeted Mass Spectrometry

Joseph Longworth and Gunnar Dittmar

Abstract

Protein homeostasis is essential for the survival of cells. It is closely related to the functioning of the ubiquitin-proteasome system, which utilizes the small protein ubiquitin as a posttranslational modifier (PTM). Clinically, the modification is of great importance as its disruption is the cause of many diseases. Unlike other PTMs, ubiquitin can encode several cellular signals by being attached as a single molecule or as a chain of several ubiquitins in various conformations. Thus, ubiquitin signaling is dependent not only on the site of attachment but also on the chain type, the so-called ubiquitin chain topology.

The most reliable quantification method for the chain topology uses a bottom-up targeted mass spectrometry based proteomics technique. While similar to other targeted proteomics techniques, the measurement of ubiquitination chain topology is complicated. First, the ubiquitin chains in the sample have to be biochemically stabilized. Second, the selection of peptides for the analysis is restricted to a given set harboring the PTMs and does not allow for optimization for amenability to mass spectrometry-based quantification. Instead, the topology-characteristic peptides are fixed. We here present such a methodology, including notes for a successful application.

Key words Ubiquitin, Chain topology, Mass spectrometry, Proteomics, Parallel reaction monitoring (PRM), Quantitative proteomics, Posttranslational modification (PTM)

1 Introduction

The identification of the ubiquitin-proteasome system as a major system for targeted degradation in 1975 [1] paved the way for studying a new kind of posttranslational modification (PTM), protein-based PTMs. The attachment of ubiquitin or one of its relatives is similar to the attachment of small molecules like phosphates in terms of reversibility (kinases vs. phosphatases) but has more degrees of freedom in terms of signaling. Ubiquitin as the founding member is integral to the maintenance of proteostasis, playing a role in among other aspects proteasomal degradation, protein stabilization, autophagy, cell cycle control, and protein trafficking [2]. It has thus been identified in a wide range of dis-

Caroline A. Evans et al. (eds.), *Mass Spectrometry of Proteins: Methods and Protocols*, Methods in Molecular Biology, vol. 1977, https://doi.org/10.1007/978-1-4939-9232-4_3, © Springer Science+Business Media, LLC, part of Springer Nature 2019

ease conditions including cancer [3, 4] and neurogenerative diseases [5, 6], as well as playing a role in the immune system [7, 8].

Differently to other PTMs, ubiquitin needs a cascade of enzymes in order to be covalently attached to the target substrate. For most ubiquitination events, the cascade consists of an activating enzyme (E1) which binds the C-terminus of ubiquitin under energy consumption, driven by hydrolyzation of one ATP to AMP. The activated ubiquitin is then transferred to an E2 enzyme, which is joined by a ligase (E3) before its C-terminus is covalently bound to the ε-amino group of a lysine in the target protein. For a review of the ubiquitination cascade *see* Hershko and Ciechanover [9].

Rather than being a single modification with a unique effect, ubiquitination can occur in a variety of forms. The simple attachment of a single ubiquitin, single ubiquitins at multiple sites or a poly-ubiquitin chain where several ubiquitins are linked to each other assembling a chain on the target protein can signal for different cellular events [10]. These chains form different topologies referred to by their linkage. They include seven linkages to the lysine residues of ubiquitin K06, K11, K27, K29, K33, K48, and K63 as well as the linear (M1) attachment [11] forming eight topologies in total. The type of chain formed is highly important as, not only can the phenotypic effect of different topologies be different, it can be contraindicative [11]. Kwon and Ciechanover provide a good current review of the associated phenotypes with each changing topology [12].

Although chain topologies can be assessed through antibody-based techniques, mass spectrometry-based targeted proteomics provides a more robust quantitative approach [10]. The typical bottom-up proteomics workflow consists of the digestion of proteins by proteases followed by liquid chromatography coupled mass spectrometry (LC-MS) analysis [13, 14]. When such a bottom-up approach with trypsin as the digesting protease is applied to ubiquitinated proteins, the cleaved C-terminal of the attached ubiquitin is left on the site of ubiquitin attachment forming a -GG modification on a lysine side chain [15]. The lysine modified by ubiquitin will also be protected from cleavage forming a peptide carrying a missed-cleavage. Thus, for the digestion of each of the eight poly-ubiquitinated chain topologies, eight topology-characteristic peptides are formed (Table 1).

These topology-characteristic peptides can be targeted for quantification. Both selective reaction monitoring (SRM) and high-resolution accurate mass (HRAM) spectrometry parallel reaction monitoring (PRM) can be utilized in data-dependent mode of MS [16]. A combination of mTRAQ chemical derivatization with SWATH (data independent) acquisition on

Table 1
Overview of the topology-characteristic peptides for each ubiquitin topology

Chain type	Topology-characteristic peptides	Most observable charge state	Observed intensity for 50 fmol/μL	ACN Elution point from C18 (relative distribution)
K06	M Q I F V K T L T G K G G	3+	8.73E+07	21% (0.625)
K11	T L T G K T I T L E V E P S D T I E N V K G G	3+	3.43E+06	26% (0.9375)
K27	T I T L E V E P S D T I E N V K A K G G	3+	2.71E+06	25% (0.875)
K29	A K I Q D K G G	2+	6.34E+05	11% (0)
K33	I Q D K E G I P P D Q Q R G G	3+	2.94E+09	16% (0.3125)
K48	L I F A G K Q L E D G R G	3+	4.41E+08	24% (0.8125)
K63	T L S D Y N I Q K E S T L H L V L R G G	4+	1.94E+06	27% (1)
Linear (M1)	G G M Q I F V K	2+	4.55E+08	25% (0.875)

The human sequence is shown for each as well as the most observable charge state, intensity observed given a loading of 50 fmol/μL on a Q-Exactive mass spectrometer and the percentage acetonitrile at which the peptide is typically eluted from C_{18}-chromatography.

a Q-TOF mass spectrometer provides a method for selective and enhanced detection and relative quantification of Ub-derived isopeptides as described in the accompanying chapter by Chicooree and Griffiths. All of these techniques involve the tracking of transitions, namely the intensity of a specified product-ion given the selection of a specified precursor ion. Typically, such transitions are scheduled based on the chromatographic separation such as to maximize the observation of a particular transition during its elution.

During typical SRM and PRM development, several peptides will be assessed for their proteotypic behavior. This is shorthand for whether the observed peptide is unique for the protein of interest and whether the peptide's intensity tracks consistently with the targeted protein intensity. With ubiquitin topology, we are however constrained to the eight topology-characteristic peptides and consideration of unfavorable characteristics for targeted proteomics must be considered. In this chapter, we present the methodology for analysis of poly-ubiquitination chain topology by PRM and some insights into the application of the technique.

2 Materials

All solvents should be HPLC grade or better.

2.1 Equipment

1. Sonication probe.
2. Cooled centrifuge.
3. Vortex.
4. Heating block or controllable incubator.
5. Vacuum concentrator.
6. HRAM LC-MS system (e.g., Q-Exactive plus, Thermo scientific).
7. Trap column containing C_{18} stationary phase with 3 μm pore particle size 0.075 mm diameter 2 cm long.
8. Analytical column containing C_{18} stationary phase with 2 μm pore particle size 0.075 mm diameter 15 cm long.

2.2 Consumables

1. Protein Lo-bind Eppendorf tube.
2. C_{18} solid phase extraction cartridges (e.g., Sep-Pak tC18, Waters).

2.3 Reagents

1. Phosphate buffered saline (PBS): 137 mM NaCl, 2.7 mM KCl, 10 mM Na_2HPO_4, 1.8 mM KH_2PO_4, pH 7.2.
2. Ubiquitin stabilization buffer: 8 M urea, 50 mM HEPES, 10 mM *N*-ethylmaleimide (has to be prepared freshly each time), pH 7.6.
3. *Escherichia coli* lysate (DH10B).
4. Dithiothreitol (DTT).
5. Chloroacetamide (CAA).
6. Endopeptidase Lys C.
7. Ammonium bicarbonate.
8. Trypsin (sequencing grade).
9. Stable isotope labeled peptides.
10. Acetonitrile (ACN).
11. Formic acid (FA).
12. Trifluoroacetic acid (TFA).

2.4 Buffers

1. Buffer A: 2% ACN, 0.1% TFA.

2.5 Software

1. Skyline (version 4.1 was used for method description though we recommend the use of Skyline-daily) [17].

3 Methodology

3.1 Sample Preparation and Lysis

1. Culture or tissue upon which ubiquitin topological analysis is to be performed (*see* **Notes 1** and **2**) is placed on ice to reduce the activity of de-ubiquitinases (*see* **Note 3**).
2. Limited sample washes can be performed with chilled PBS (*see* **Note 2**).
3. Resuspend sample in ubiquitin stabilization buffer.
4. Lyse cells according to the cell type used. Typically, cell lines can be lysed using a sonication probe. Using a Branson sonifier SFX 150 equipment with a micro tip pulse each sample for five rounds of 10 s on 10 s off while on ice at 70% intensity (for other sonifiers required intensity may differ).
5. Centrifuge each sample at 18,000 × *g* for 10 min at 5 °C.
6. Transfer the supernatant containing suspended proteins to a fresh tube and store at −80 °C until ready for processing.

3.2 In-Solution Sample Processing (*See* Note 4)

1. To each sample to be processed add 0.5 μg of *E. coli* lysate (*see* **Note 5**).
2. Reduce the sample by addition of 10× DTT giving a final concentration in the sample of 10 mM. Briefly mix using a vortex before incubating at 37 °C for 30 min.
3. Alkylate the sample by addition of 10× CAA giving a final concentration of 55 mM. Incubate at room temp in the dark for 20 min (*see* **Note 6**).
4. Add Lys C to the sample at a 1:25 (w/w) ratio. Incubate at 37 °C for 5 h (*see* **Note 7**).
5. Dilute the samples with 3× the current sample volume with 50 mM ammonium bicarbonate.
6. Add trypsin to the sample at a 1:25 (w/w) ratio. Incubate at 37 °C for 10 h (*see* **Note 8**).
7. Dry samples on a vacuum concentrator.

3.3 Spike Samples with Heavy Labeled Counterparts and C18 Desalting

1. Heavy peptides mirror the peptide of interest in a way that they have the same amino acids sequence, but the last lysine or arginine residue is replaced by an isotopically heavy lysine or arginine. The -GG modification must also be included on the heavy peptide (*see* **Notes 9** and **10**).
2. Prepare a mix of the heavy peptides to be added to the sample in buffer A (*see* **Note 11**).
3. Resuspend the digested and dried sample using the prepared heavy peptide mix.
4. Desalt using a C_{18} cleanup cartridge following manufactures instructions.

5. Post elution from the desalting column, dry samples in a vacuum concentrator.
6. Resuspend the peptides in buffer A to get them ready for LC-MS analysis.

3.4 LC-MS Analysis

1. Set up LC parameters with flow rates appropriate to your column manufacturer and LC-MS instrument. Load the sample onto a trapping column allowing for some time to wash unbound material off the trapping column before performing a valve switch bringing the trapping column in line with the analytical column and the mass spectrometer. Use a linear gradient going from low to high organic to achieve a good chromatographic separation (*see* **Note 12**).
2. Set up the MS parameters for a PRM analysis (*see* **Note 13**). It is good to intersperse PRM experiments with an MS full scan within an LC-MS run to record a total ion chromatogram (TIC).
3. Upload the inclusion list either scheduled or unscheduled to the mass spectrometer as described in Subheading 3.5 below.
4. Inject ~1 μg of the sample onto an LC-MS system.

3.5 Scheduling Determination

1. A scheduling run should be performed before sample analysis. By preference, this should be done with the heavy peptides with carrier peptides as discussed in **Note 4**.
2. Create a new Skyline document (*see* **Note 14**).
3. Click on Edit > Insert > Peptides.
4. Input the peptide sequence for each of the topology-characteristic peptides (human variants are shown in Table 1; *see* **Note 15**).
5. Label each peptide with a protein name referring to the associated topology and select "Insert Peptides."
6. Right-click on the peptide sequence in the "Targets" pane, selecting Modify. Add a -GG modification to the amino acid K with the chemical formula $H_6C_4N_2O_2$.
7. Click on Settings > Peptide Settings and navigate to the Modifications tab to input a heavy label as associated with the synthetic peptide purchased for spiking into the sample.
8. Right-click on each sequence and select pick children to select the charge state from which to perform the PRM.
9. Derive an isolation list by clicking File > Export > Isolation List.
10. Select the instrument being used.
11. For initial scheduling, use a standard method type. Once a scheduling has been performed, use a scheduled method type derived from a recent run by clicking OK.

3.6 Data Interpretation

1. Upon run completion, data can be analyzed through use of Skyline. Load a file with the topology-characteristic peptides annotated as described in Subheading 3.5.
2. Click on File > Import > Results.
3. Click OK and select the LC-MS data to be processed.
4. Using the heavy standard identify the transitions giving the superior signal to noise and fragmentation pattern similarity.
5. Select the boundaries for peak integration (*see* **Note 16**).
6. Relative peak intensities can be observed by clicking on View > Peak Areas > Replicate Comparison.
7. Raw data can be exported by clicking File > Export > Report.

4 Notes

1. Upon digestion, the ubiquitination state of the attached ubiquitin is decoupled. Thus, it is not possible to deduce the purity of chain for a single linkage type, or the association of linkage with its target protein. Thus, mass spectrometer-based ubiquitin topology analysis is limited to quantification of the global chain topology profile of a sample. This can be an entire extract, an isolated fraction of a sample such as an organelle extraction, or an enrichment of a protein of interest such as a pull-down.
2. Any sample derivation must take into consideration the propensity for de-ubiquitination and chain-breakdown. De-ubiquitinases are highly efficient and ubiquitin chains can easily be broken down during sample preparation even in denaturing conditions like 2 M urea.
3. A positive control for analysis can be constructed by treatment of cells with protease inhibitors (e.g., 10 mM MG-132 for 1 h prior to harvest). By blocking the proteasome the accumulation of different chains, primarily K48-chains, is induced.
4. Here we present an in-solution methodology. An in-gel methodology is also applicable although the use of a combination of Lys C and trypsin as a single digestion step is recommended to promote digestion of the digestion-resistant ubiquitin.
5. The topology-characteristic peptides are to varying degrees adherent to plasticware and glassware used during sample processing. Figure 1 shows the effect of increasing levels of *E. coli* carrier protein on the various topology-characteristic peptides. Consideration of varying levels of background peptides of the sample must, therefore, be made. This issue can be aided by the addition of an *E. coli* lysate acting as carrier protein/peptide.

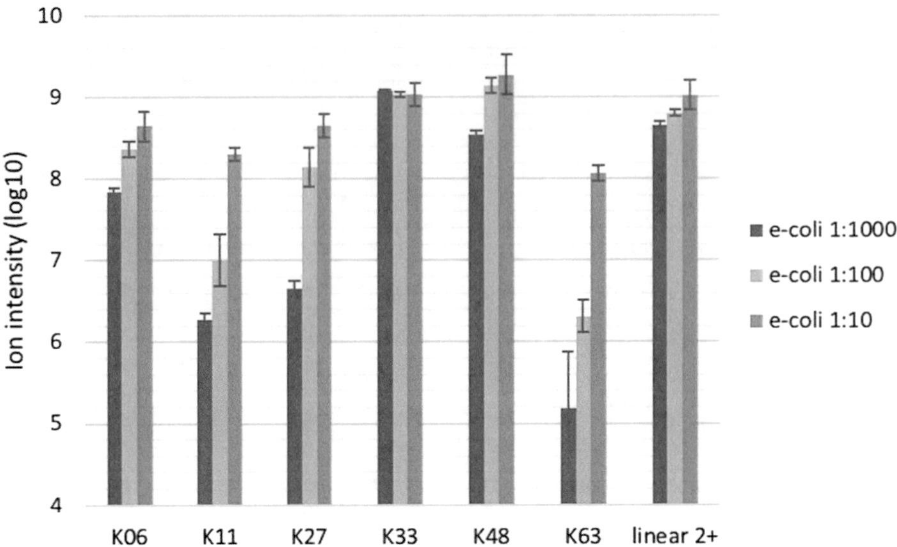

Fig. 1 Comparison of carrier protein. The observed ion intensity for each ubiquitin topology is shown with increasing levels of *E. coli* carrier protein. The effect of a carrier protein has a significant increase in K63, K27, and K11 while it is less influential on Linear and K33

6. CAA is used rather than iodoacetamide (IAA). IAA can readily form a double modification of Lysine residues with an additional mass of 114 Da. This is indistinguishable from the GG modification [18].
7. In-solution digestion is performed in two stages. Initial digestion is done with Lys C which remains active at the 8 M urea condition used to protect ubiquitin chains as well as promoting ubiquitin denaturation of the relatively digestion-resistant ubiquitin protein. For a robust quantification, a complete digest of ubiquitin is essential.
8. The ratio of Lys C and trypsin is higher than typical in bottom-up proteomics to promote digestion of ubiquitin, a digestion-resistant protein.
9. Though it is possible to identify and perform label-free quantification of the ubiquitin chains it is preferable to use heavy labeled peptides. This allows for a run-to-run calibration and absolute quantification (absolute if using high accuracy synthetic peptides such as AQUA. Otherwise relative quantification) of the topology-characteristic peptide and thus the level of each ubiquitin chain topology.
10. Heavy peptides should be kept at high concentrations and in high organic conditions to prevent loss of peptide due to adhesion to plasticware. This is particularly important for K63 which readily adheres even to lo-bind plasticware.
11. Ensure dilution of the organic conditions used for heavy peptide storage to below 5% by dilution in an appropriate loading buffer (2% ACN, 0.1% TFA).

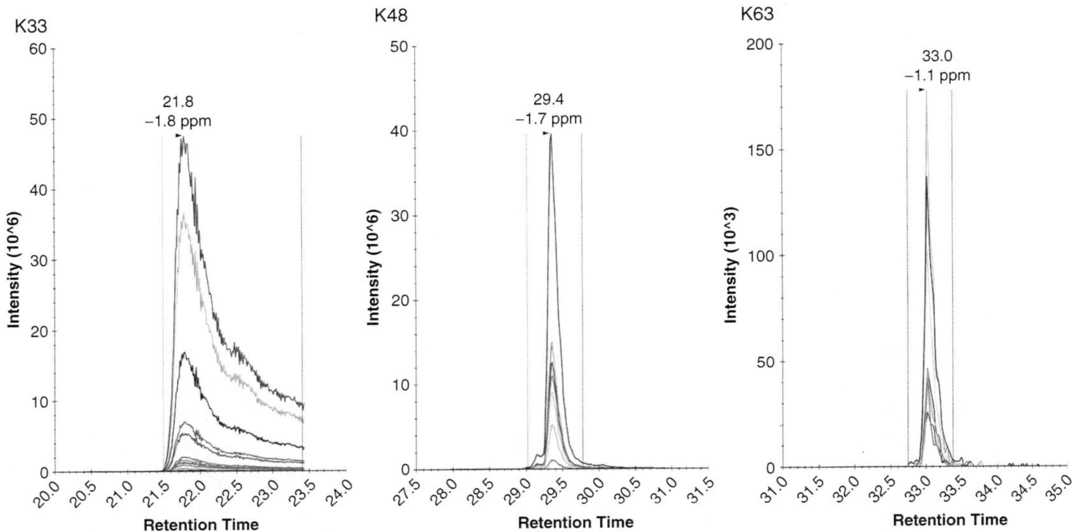

Fig. 2 Extracted ion chromatogram showing the typical elution profile of K33, K48, and K63 ubiquitin topology

12. A suitable chromatographic gradient for this analysis would be a linear gradient exchanging buffer A (MS grade water, 0.1% FA) for buffer B (ACN, 0.1% FA) from 2% to 35% at 1% per minute. To be followed by an increase to 90% buffer B for 8 column volumes to clean the column and 22 column volumes at 2% buffer B to recondition the column.
13. Appropriate settings for a Q-Exactive HF would be a resolution of 70,000, AGC target of 1e6, Maximum IT of 240 ms for the survey scan with a resolution of 35,000, AGC target of 1e6. Maximum IT of 150 ms, Loop count of 100 and isolation window of 1.0 *m/z*.
14. There is alternative software available for PRM scheduling and interpretation. We here show a workflow with Skyline as it is widely used, open source and readily compatible with all mass spectrometry vendors.
15. In addition, four tryptic peptides greater than seven amino acids are formed from unmodified ubiquitin. For human, they consist of (TITLEVEPSDTIENVK, TLSDYNIQK, ESTLHLVLR, EGIPPDQQR). EGIPPDQQR does not elute with a clean peak while TITLEVEPSDTIENVK does not have good quantitative properties hence TLSDYNIQK, ESTLHLVLR are recommended for total ubiquitin quantification. This should be done with consideration of the reduction of such peptides with incorporation into K63 chains.
16. As can be seen from the example chromatograms in Fig. 2, the K33 peptide does not elute cleanly given reverse phase chromatography. It is, however, reproducible and can thus be used for quantitation given a consistent selection of the range used to interpret quantification of the signal.

Acknowledgments

The authors would like to thank Dr. Antoine Lesur for his helpful advice on various technical issues examined in this chapter and reviewing the manuscript.

References

1. Goldstein G, Scheid M, Hammerling U et al (1975) Isolation of a polypeptide that has lymphocyte-differentiating properties and is probably represented universally in living cells. Proc Natl Acad Sci U S A 72:11–15
2. Pickart CM, Eddins MJ (2004) Ubiquitin: structures, functions, mechanisms. Biochim Biophys Acta 1695:55–72
3. Diehl JA, Fuchs SY, Haines DS (2010) Ubiquitin and cancer. Genes Cancer 1:679–680
4. Senft D, Qi J, Ronai ZA (2018) Ubiquitin ligases in oncogenic transformation and cancer therapy. Nat Rev Cancer 18:69–88
5. Zheng Q, Huang T, Zhang L et al (2016) Dysregulation of ubiquitin-proteasome system in neurodegenerative diseases. Front Aging Neurosci 8:303
6. Atkin G, Paulson H (2014) Ubiquitin pathways in neurodegenerative disease. Front Mol Neurosci 7:63
7. Zinngrebe J, Montinaro A, Peltzer N, Walczak H (2014) Ubiquitin in the immune system. EMBO Rep 15:28–45
8. Hu H (2016) Ubiquitin signaling in immune responses. Cell Res 26:27
9. Hershko A, Ciechanover A (1998) The ubiquitin system. Annu Rev Biochem 67:425–479
10. Beaudette P, Popp O, Dittmar G (2016) Proteomic techniques to probe the ubiquitin landscape. Proteomics 16:273–287
11. Akutsu M, Dikic I, Bremm A (2016) Ubiquitin chain diversity at a glance. J Cell Sci 129: 875–880
12. Kwon YT, Ciechanover A (2017) The ubiquitin code in the ubiquitin-proteasome system and autophagy. Trends Biochem Sci 42: 873–886
13. Aebersold R, Mann M (2003) Mass spectrometry-based proteomics. Nature 422: 198–207
14. Chait BT (2006) Mass spectrometry: bottom-up or top-down? Science 314:65–66
15. Goldknopf IL, Busch H (1977) Isopeptide linkage between nonhistone and histone 2A polypeptides of chromosomal conjugate-protein A24. Proc Natl Acad Sci U S A 74:864–868
16. Lesur A, Domon B (2015) Advances in high-resolution accurate mass spectrometry application to targeted proteomics. Proteomics 15:880–890
17. MacLean B, Tomazela DM, Shulman N et al (2010) Skyline: an open source document editor for creating and analyzing targeted proteomics experiments. Bioinformatics 26: 966–968
18. Nielsen ML, Vermeulen M, Bonaldi T et al (2008) Iodoacetamide-induced artifact mimics ubiquitination in mass spectrometry. Nat Methods 5:459–460

Chapter 4

Quantitative Phosphoproteomic Using Titanium Dioxide Micro-Columns and Label-Free Quantitation

Martin E. Barrios-Llerena and Thierry Le Bihan

Abstract

Phosphorylation events are important during cellular function. Analysis of phosphorylation in complex samples has been extensively studied using large-scale phosphopeptide enrichment methods. Quantitative analysis of the enriched phosphopeptides is subsequently performed using label-based methodologies (e.g., SILAC, iTRAQ, and others). Here we describe the protocol for the quantitative analysis of phosphopeptides, enriched with titanium dioxide micro-column, using an intensity-based label-free quantitation.

Key words Phosphopeptide enrichment, Titanium dioxide chromatography (TiO_2), Label-free quantitation, Mass spectrometry

1 Introduction

Phosphorylation is a reversible posttranslational protein modification (PTM) that has been extensively studied. This PTM is critical for the regulation of a wide spectrum of biological events and cellular functions such as signal transduction, gene expression, cell progression, metabolism, growth, and disease [1]. Identification and quantification of the phosphorylation state of proteins is an essential step in the continued elucidation of cellular function.

Phosphoproteome characterization is challenging due to the low abundance/stoichiometry of the phosphorylation sites in cells/tissue. Therefore, enrichment is a requirement to perform phosphoproteome analysis. Several methodologies have been used for the enrichment of phosphopeptides, especially using TiO_2 [2, 3] or Immobilized Metal Affinity Chromatography (IMAC) [4].

Phosphorylation dynamics, assessed via relative quantification, have been performed using label-based methodologies such as stable isotope labeling by amino acid in cell culture (SILAC) [5, 6], metabolic labeling with stable isotopes such as nitrogen (15N) or (18O) [7, 8], or by introducing in vitro stable isotope-coded reagents (e.g., iTRAQ, TMT) [9–11]. These approaches are

Caroline A. Evans et al. (eds.), *Mass Spectrometry of Proteins: Methods and Protocols*, Methods in Molecular Biology, vol. 1977, https://doi.org/10.1007/978-1-4939-9232-4_4, © Springer Science+Business Media, LLC, part of Springer Nature 2019

limited to a binary, ternary, or multiplexing (up to ten samples) comparisons, further limiting the breadth and depth of phosphopeptide analysis.

A more straightforward approach is a label-free quantification, which alleviates the restriction on the number of samples that can be quantitatively monitored, and can reach a very high dynamic range in quantification.

The methodology described in this chapter makes use of TiO_2 for phosphopeptide enrichment [2, 12, 13] in a micro-column format, and high-resolution mass spectrometry for phosphopeptides identification. Quantification of the enriched phosphopeptides was carried out using an intensity-based label-free quantitation approach.

2 Materials

2.1 Protein Extraction and Digestion

1. Extraction Buffer: 8 M urea in 200 mM HEPES (4-(2-hydroxyethyl)-1-piperazineethanesulfonic acid) buffer to pH 8.0 (*see* **Note 1**).
2. Reduction solution: 0.5 M *tris*(hydroxypropyl)phosphine (THP) solution (*see* **Note 2**). Store at −20 °C.
3. Alkylating solution: 0.5 M iodoacetamide solution (*see* **Note 3**).
4. 1 M ammonium bicarbonate (ABC).
5. Formic Acid.
6. Trypsin (modified, sequencing grade).
7. SPE cartridge, 100 mg C18 packing material, SupelCleanC18 cartridge (*see* **Note 4**).

2.2 Titanium Dioxide (TiO_2) Chromatography

1. Titanium dioxide (TiO_2) beads, 10 μm (Titansphere, GL Sciences, Japan).
2. PEEK tubing, 4 cm × 1/16″ × 0.040″, grey.
3. P-627Adapter, 10–32 F to 1/4–28 F PEEK (*see* **Note 5**).
4. LT-115—Super Flangeless™ Nut, Short, PEEK, 1/4–28 Flat-Bottom, for 1/16″—1/32″ OD.
5. Frit in Ferrule 0.5 μm SF PEEK Blk w/SS Ring—P-275x.
6. Peptide loading buffer: 2.5% acetonitrile (ACN) and 0.5% trifluoroacetic acid (TFA).
7. TiO_2 washing buffer 1: 80% ACN, 200 mg/mL 2,5-dihydroxybenzoic acid (DHB).
8. TiO_2 washing buffer 2: 0.5% TFA, 80% ACN.
9. TiO_2 washing buffer 3: 0.1% TFA, 80% ACN.

10. TiO_2 elution buffer 1: 400 mM ammonium solution.
11. TiO_2 elution buffer 2: 5% (v/v) ammonium in water.
12. TiO_2 elution buffer 3: 5% (v/v) ammonium in ACN.

2.3 Phosphopeptide Quantitation

1. MASCOT Version 2.4 (Matrix Science Ltd, UK).
2. Progenesis 4.0 (Nonlinear Dynamics, UK).
3. Maxquant software.

3 Methods

3.1 Peptide Preparation for TiO_2 Enrichment

1. Cell/tissue, collected according to laboratory protocols, was lysed by adding Extraction Buffer, cell/tissue was disrupted by sonication.
2. Protein extract (300 μg) was reduced in 5 mM THP at room temperature for 15 min. After reduction, cysteines were alkylated in 10 mM iodoacetamide at room temperature for 30 min in the dark. After incubation, the reaction was quenched with 5 mM THP for 15 min at room temperature.
3. Before digestion with trypsin, the urea concentration was diluted down to ~1.5 M urea with Milli-Q water and ammonium bicarbonate to 25 mM ABC, and trypsin was added to the sample at a ratio of 1:50 enzyme/protein and allowed to digest overnight at 37 °C.
4. After tryptic digestion, formic acid was added to 2% final concentration to stop the reaction. Peptides were cleaned up using a SupelCleanC18 cartridge and dried under low pressure in a Speed Vac. Peptide samples were stored at −20 °C.

3.2 Packing the TiO_2 Micro-Column

Assembling of the TiO_2 micro-column is depicted in Fig. 1, and explained below:

1. Attach the frit in ferrule PEEK to one end of the 4-cm PEEK tubing.
2. Assemble this end to a flat-bottom flangeless Nut (LT-115) with 10–32 F to 1/4–28 F PEEK adapter (P-627).
3. To the other end of this assembly attach a small piece of gray PEEK tubing with a PEEK finger tight nut (F-300X, winged).
4. Suspend ~2 mg TiO_2 beads in 500 μL 100% methanol or isopropanol.
5. Prepare a TiO_2 micro-column by packing the TiO_2 beads into the PEEK tubing using a pressure pump.
6. After the tubing is fully packed with the TiO_2 beads close this end with the frit in ferrule PEEK in a similar manner as in **steps 1** and **2**, column is ready to be used (*see* **Note 6**).

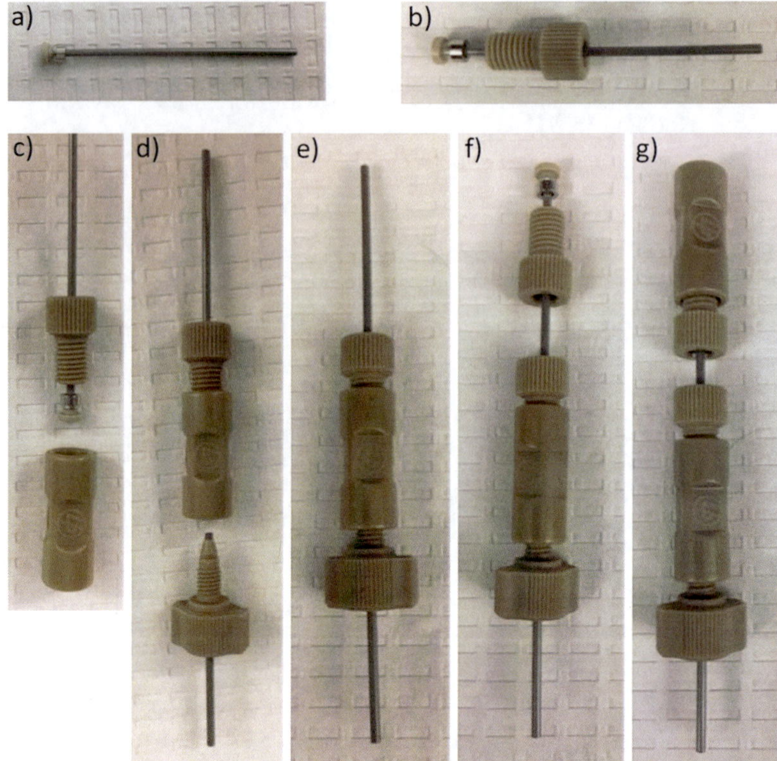

Fig. 1 Step by step assembly of the TiO_2 micro-column. (**a**) Attach the Frit in a ferrule together with the SS ring to the grey PEEK tubing; (**b**) insert the Super Flangeless™ Nut (LT-115) to the previous assembly; (**c**) attach this assembly to the Adapter (P-627); (**d**, **e**) to the other end of the assembly insert a piece of grey PEEK tubing with a 1/16″ nut; (**f**) after packing the micro-column with the TiO_2 particles, insert a Super Flangeless™ Nut and the Frit in a ferrule together with the SS ring to the grey PEEK tubing as in step (**a**); (**g**) finally attach the Adapter (P-627) to this. The micro-column is ready to be used for phospho-peptide enrichment

3.3 Loading Peptides onto the TiO_2 Micro-Column

1. Wash the Micro-column, first with 200 μL of Peptide loading buffer followed by a wash with TiO_2 washing buffer 2.
2. Dried peptides were dissolved in equal volume of Peptide loading buffer and TiO2 washing buffer 1 (100 μL each).
3. Peptides were loaded to the Micro-column using a pressure vessel.
4. Micro-column was washed first with 200 μL of Loading peptide buffer followed by 400 μL of TiO_2 washing buffer 1.
5. A further wash with 400 μL of TiO_2 washing buffer 2 and 600 μL of TiO_2 washing buffer 3 was performed.

3.4 Elution of Phosphorylated Peptides from the TiO_2 Micro-Column

1. Elute phosphopeptides from the TiO_2 micro-column using 50 μL of TiO_2 elution buffer 1, followed by a second elution with 50 μL of TiO_2 elution buffer 2.
2. Finally, elute the strongly bound phosphopeptides with 50 μL of TiO_2 elution buffer 3. Pool the three elution and dry under low pressure. Store dried phosphopeptides at −20 °C.

3.5 Data Analysis and Quantitation

1. Dried phosphopeptides eluted from the TiO_2 micro-column were resuspended in mobile phase A (2.5% acetonitrile, 0.1% formic acid) and nano-LC MS/MS analyses were performed as described previously [14] (*see* **Note 7**).
2. Manual inspection of the data set was performed. This is a good practice to ensure the quality of the phospho enrichment experiment carried out (*see* **Note 8**).
3. Label-free quantitation was performed using Progenesis 4.0 (Nonlinear Dynamics, UK) as follows:
   (a) MS spectra data acquired for each samples were loaded to Progenesis LC-MS. Profile data of the MS scans were transformed to peak lists with respective peak m/z values, intensities, abundances, and m/z width. MS/MS spectra were treated similarly.
   (b) For retention time alignment of the samples, the most complex sample was selected as a reference, and the retention times of the other samples were aligned automatically to a maximal overlay of all features (*see* **Note 9**).
   (c) Features with only one charge or more than four charges were excluded from further analyses.
   (d) Raw abundances of the remaining features were normalized to allow correction for factors resulting from experimental variation. The default method consists of using a median-adjusted approach where the median of each sample's intensity is normalized against a median intensity of the reference sample.
   (e) Rank 1–5 MS/MS spectra were exported as Mascot generic file and used for peptide identification with MASCOT Version 2.4 (Matrix Science Ltd, UK).
   (f) Search parameters were peptide mass tolerance of 7 ppm, and MS/MS tolerance of 0.4 amu allowing a maximum of two missed cleavage. Carbamidomethylation of cysteine was set as a fixed modification, and oxidation of methionine, N-terminal protein acetylation and phosphorylation at serine, threonine and tyrosine were allowed as variable modification.

(g) Peptide assignments with an ion score cutoff of 20 and a significance threshold of $p < 0.05$ were reimported to Progenesis.

(h) All multicharged ions (2+, 3+, 4+) were extracted and ion intensities summed for normalization. Peptide abundances were mean-normalized and arcsinh transformed. Within-group means were calculated to determine fold changes, and *p*-values were calculated using Progenesis (*see* **Note 10**).

4. Similar phospho-motif can be summed in order to reduce the complexity of the analysis using qpMerged [15] (a specific phospho motif which is found in a peptide of different charge states, or oxidation state or resulting from partial protease digestion).
5. For over- and underrepresentation of peptide–motifs we required a two-tailed *t*-test *p*-value of <0.05 and an absolute fold change ratio of >1.5. For unchanging motifs we required a two one-sided test (TOST) with a *p*-value of <0.05 and an absolute fold change ratio of <1.5. Global *p*-value evaluation was performed using Progenesis, and pair-wise *t*-tests were performed in *R* (3.1.0).

4 Notes

1. Urea should be added to the HEPES buffers just prior to use for protein extraction.
2. THP is more resistant to ambient oxidation and is a more effective reducing reagent.
3. Adding 50% acetonitrile to dissolving buffer helps to solubilize the IAA, and it is not detrimental to the tryptic digestion.
4. We found that these cartridges are efficient on the recovery of peptides from diluted solutions.
5. Fittings and connections used for micro-column assembly are described according to standard denomination used in liquid chromatography, and these are currently the most suitable parts for proper micro-column assembly.
6. The material packed in this micro-column is sufficient for phosphoproteomics enrichment of 200–300 μg of tryptic peptides.
7. As a rule of thumb the optimum amount of phosphorylated peptides found are around 40% of all peptide identified. Increasing this ratio can be easily achieved by increasing either the total protein concentration, used for enrichment, or the wash volume of the modifier (DHB solution). The purity will

increase, albeit to the expense of a global lower amount of phosphopeptide detected. In addition, having a core set of non-phosphorylated peptide allows for a more robust data normalization strategy.

8. Most of the phosphopeptide should be serine followed by threonine and tyrosine in the rough proportion (1000:100:1). As a dataset can be quite large, concentrating on the pool of significantly differentially expressed peptide is a good compromise. Under HCD and CID it is common to observe a neutral lost. Phosphoserine based peptides should be among the most intense peak in the spectra (MH-H3PO4). For a 2+ charged peptide one should find an intense peak around 48.9884 amu of lower mass than the precursor ion. If such an ion is not found, looking for multiple lost is the next step (MH-H_3PO_4-NH_3 or MH-H_3PO_4-H_2O).
9. Another practical way to ensure maximal overlay of the features is using a pooled sample, which contain features from most of the different group.
10. Although quite stringent approach, for large dataset, it is essential to correct *p*-value for multiple comparisons. For example, both the Holm–Bonferroni and the Šidák methods are good method, which can be slightly less stringent than the Bonferroni correction.

References

1. Graves JD, Krebs EG (1999) Protein phosphorylation and signal transduction. Pharmacol Ther 82(2–3):111–121
2. Larsen MR, Thingholm TE, Jensen ON et al (2005) Highly selective enrichment of phosphorylated peptides from peptide mixtures using titanium dioxide microcolumns. Mol Cell Proteomics 4(7):873–886
3. Thingholm TE, Jorgensen TJ, Jensen ON et al (2006) Highly selective enrichment of phosphorylated peptides using titanium dioxide. Nat Protoc 1(4):1929–1935
4. Cantin GT, Yi W, Lu B et al (2008) Combining protein-based IMAC, peptide-based IMAC, and MudPIT for efficient phosphoproteomic analysis. J Proteome Res 7(3):1346–1351
5. Ong SE, Blagoev B, Kratchmarova I et al (2002) Stable isotope labeling by amino acids in cell culture, SILAC, as a simple and accurate approach to expression proteomics. Mol Cell Proteomics 1(5):376–386
6. Altelaar AF, Frese CK, Preisinger C et al (2013) Benchmarking stable isotope labeling based quantitative proteomics. J Proteome 88:14–26
7. Kwon OK, Kim SJ, Lee YM et al (2016) Global analysis of phosphoproteome dynamics in embryonic development of zebrafish (Danio rerio). Proteomics 16(1):136–149
8. Cantin GT, Venable JD, Cociorva D et al (2006) Quantitative phosphoproteomic analysis of the tumor necrosis factor pathway. J Proteome Res 5(1):127–134
9. Hu X, Wu L, Zhao F et al (2015) Phosphoproteomic analysis of the response of maize leaves to drought, heat and their combination stress. Front Plant Sci 6:298
10. Nguyen TH, Brechenmacher L, Aldrich JT et al (2012) Quantitative phosphoproteomic analysis of soybean root hairs inoculated with Bradyrhizobium japonicum. Mol Cell Proteomics 11(11):1140–1155
11. Ma Q, Wu M, Pei W et al (2014) Quantitative phosphoproteomic profiling of fiber differentiation and initiation in a fiberless mutant of cotton. BMC Genomics 15:466
12. Jensen SS, Larsen MR (2007) Evaluation of the impact of some experimental procedures on different phosphopeptide enrichment tech-

niques. Rapid Commun Mass Spectrom 21(22):3635–3645

13. Thingholm TE, Larsen MR (2009) The use of titanium dioxide micro-columns to selectively isolate phosphopeptides from proteolytic digests. Methods Mol Biol 527:57–66. xi
14. Le Bihan T, Hindle M, Martin SF et al (2015) Label-free quantitative analysis of the casein kinase 2-responsive phosphoproteome of the marine minimal model species Ostreococcus tauri. Proteomics 15(23–24):4135–4144
15. Hindle MM, Le Bihan T, Krahmer J et al (2016) qpMerge: merging different peptide isoforms using a motif centric strategy. bioRxiv 047100; https://doi.org/10.1101/047100

Chapter 5

Isotopic Labeling and Quantitative Proteomics of Acetylation on Histones and Beyond

Peder J. Lund, Yekaterina Kori, Xiaolu Zhao, Simone Sidoli, Zuo-Fei Yuan, and Benjamin A. Garcia

Abstract

Lysine acetylation is an important posttranslational modification (PTM) that regulates the function of proteins by affecting their localization, stability, binding, and enzymatic activity. Aberrant acetylation patterns have been observed in numerous diseases, most notably cancer, which has spurred the development of potential therapeutics that target acetylation pathways. Mass spectrometry (MS) has become the most adopted tool not only for the qualitative identification of acetylation sites but also for their large-scale quantification. By using heavy isotope labeling in cell culture combined with MS, it is now possible to accurately quantify newly synthesized acetyl groups and other PTMs, allowing differentiation between dynamically regulated and steady-state modifications. Here, we describe MS-based protocols to identify acetylation sites and quantify acetylation rates on both proteins in general and in the special case of histones. In the experimental approach for the former, ^{13}C-glucose and D_3-acetate are used to metabolically label protein acetylation in cells with stable isotopes, thus allowing isotope incorporation to be tracked over time. After protein extraction and digestion, acetylated peptides are enriched via immunoprecipitation and then analyzed by MS. For histones, a similar metabolic labeling approach is performed, followed by acid extraction, derivatization with propionic anhydride, and trypsin digestion prior to MS analysis. The procedures presented may be adapted to investigate acetylation dynamics in a broad range of experimental contexts, including different cell types and stimulation conditions.

Key words Histone acetylation, Protein acetylation, Epigenetics, Mass spectrometry, Acetylation dynamics, Proteomics, Isotopic labeling, Posttranslational modifications

1 Introduction

Posttranslational modifications (PTMs), which can bestow different biophysical properties to otherwise identical sequences of amino acids, serve as a primary means by which protein function is regulated. While this paper focuses on methods dealing with the analysis of acetylation at lysine residues, especially the quantitative analysis of acetylation dynamics, acetylation represents just one of the many types of PTMs known to occur on proteins. Nonetheless,

Caroline A. Evans et al. (eds.), *Mass Spectrometry of Proteins: Methods and Protocols*, Methods in Molecular Biology, vol. 1977, https://doi.org/10.1007/978-1-4939-9232-4_5, © Springer Science+Business Media, LLC, part of Springer Nature 2019

much is known about lysine acetylation and its effect on protein stability, localization, and function in the broader context of regulating cell signaling pathways and gene expression. Acetylation of lysine residues occurs through the catalytic activity of lysine acetyltransferases (KAT), which transfer the acetyl moiety from acetyl-CoA to the ε-amino group of a specific lysine residue [1]. Deacetylases (KDAC) mediate the removal of acetylation through a mechanism dependent on Zn^{2+} or NAD^{+} [1, 2], making this PTM dynamic and reversible.

The first observation of protein acetylation occurred in the 1960s on histones [3, 4], which provided a foundation for the modern field of epigenetics as it relates to histone modifications. In eukaryotic cells, histones associate closely with DNA to form the basic repeating unit of chromatin, the nucleosome, which consists of approximately 147 bp of DNA wrapped around a histone octamer composed of the core histones H2A, H2B, H3, and H4 [5, 6]. In addition to providing a structural scaffold that permits packaging of DNA in cells, histone proteins have a major role in the regulation of genetic processes, such as DNA replication, DNA repair, and transcription [7, 8]. In many cases, this regulation involves the N-terminal tails of histones, which protrude out from the nucleosome and are subject to a multitude of PTMs, including acetylation, methylation, phosphorylation, ubiquitination, and ADP ribosylation [2, 9–11]. In the contemporary paradigm of histone modifications, these marks, which are thought to represent a combinatorial "histone code" [12], are placed by "writers," recognized by "readers," and removed by "erasers" [13]. Since the topic at hand is protein acetylation, further discussion is focused specifically on histone acetylation, which usually correlates with the transcriptional activation of a genomic locus. Histone acetylation is "written" by histone acetyltransferases (HATs), of which three main families exist: GNAT, MYST, and CBP/p300 [2, 8, 9]. Compared to other histone modifying enzymes, histone acetyltransferases generally lack strict site selectivity, as exemplified by p300, which targets H3K14, H3K18, H4K5, and H4K8 [8, 11]. Responding to the actions of histone acetyltransferases, "reader" proteins, such as PCAF and SMARCA4 [14, 15], can then bind to the acetylated lysine residues via their bromodomains, thereby promoting additional modifications, transcription, or other biological processes. Finally, histone deacetylases (HDACs), which belong to one of four classes (I–IV), act as "erasers" of acetylation to return histones to an unmodified state [2, 9]. Beyond serving as a mechanism for the recruitment of regulatory proteins, histone modifications are also thought to affect nucleosome structure. For example, compared to unmodified lysine residues, which carry a net positive charge at physiological pH, acetylated lysine residues are neutral. Thus, acetylation may weaken electrostatic interactions between histones and the DNA, possibly affording greater accessibility to the transcriptional machinery [9, 16].

Beyond histones, acetylation has been detected on more than 80 transcription factors as well as many other nuclear, cytoplasmic, and mitochondrial proteins with significant effects on protein function [17–19]. For example, acetylation has been found to increase the stability of some proteins, such as the tumor suppressor p53, the transcriptional activator HNF6, and the ATP-dependent helicase WRN, but decrease the stability of others, such as the transcription factor GATA1. Specifically, treatment with the deacetylase inhibitor trichostatin A (TSA) not only increased the levels of acetylated p53, as expected, but also increased the half-life of p53, suggesting that acetylated p53 is more stable [20]. Similarly, for HNF6, increased levels of the acetyltransferase CBP led to increased acetylation of HNF6 with a concomitant increase in its half-life [21]. Acetylation of WRN at residues K366, K887, K1117, K1127, K1389, and K1413 by CBP was shown to suppress its ubiquitination, thus increasing its stability [22]. On the other hand, acetylation of GATA1 appears to signal its ubiquitination and subsequent degradation. TSA treatment resulted in decreased levels of GATA1, which was recovered via treatment with the proteasome inhibitor MG-132 [23]. Acetylation can also modulate, in either positive or negative fashion, the ability of a protein to interact with binding partners, as in the case of the transcriptional activator STAT3, the structural protein actin, and the molecular chaperone HSP90 [24]. Acetylation can also affect the localization of proteins. For instance, acetylated PCAF and SRY localize to the nucleus, whereas the acetylated p300 localizes to the cytosol [24]. Autoacetylation of lysine residues within the nuclear localization signal (NLS) of acetyltransferase PCAF is required for PCAF nuclear localization, potentially by affecting its interaction with nuclear import factors or by driving conformational changes [25]. Acetylation of K136 on SRY, a transcriptional regulator involved in male development, increases the SRY-importin β interaction threefold, resulting in nuclear localization of SRY [26]. Given that acetylation is important in regulating protein activity, abnormal acetylation can perturb the normal functioning of biological processes and become pathological. In particular, aberrant acetylation can lead to disease states such as tumorigenesis, cancer cell proliferation, persistence of DNA damage due to misregulation of repair machinery, dysregulation of the immune response, and neurodegeneration [1, 24, 27, 28]. In fact, bioinformatics studies on cancers show that mutations may arise more frequently at acetylation sites compared to sequences of amino acids without PTMs [29]. Although numerous studies have focused on mapping acetylation sites across the proteome [19, 30], it is also critical to consider the dynamics of acetylation at these sites. Studying acetylation dynamics can help improve understanding of the kinetics, usage, recycling, and regulation of this modification, while also providing insights about protein function. Specifically, the dynamics of acetylation indicate which sites are added or removed quickly, thereby providing information about the activity of the KATs or KDACs responsible for regulating these sites. In addition, knowledge of acetylation dynamics for a particular protein may give insight as to the role of the protein in its signaling pathway, such as whether the protein is part of a transient or more sustained signaling network. Since acetylation can also affect protein-protein interactions, knowledge of the dynamics of these sites may reveal the duration over which the protein interacts with its partners, perhaps to accomplish a given function. Acetylation on some proteins is also known to affect protein stability and localization, implying that characterization of the dynamics of the acetylation sites may provide hints about protein half-life or whether a protein spends more time in one compartment of the cell compared to another.

Protein-modifying enzymes rely on small molecule co-factors as donors of the chemical groups being transferred to proteins and in some cases (e.g., some demethylases and deacetylases), as essential components in the reaction mechanism responsible for their removal [2, 31]. Since the levels of these co-factors reflect the activity of various metabolic pathways, it is now appreciated that cell metabolism interfaces with cell signaling through PTMs [32, 33]. As mentioned, protein acetylation originates from acetyl-CoA, which bridges both glycolysis and fatty acid oxidation to the tricarboxylic acid (TCA) cycle for oxidative phosphorylation and the generation of ATP [31, 34]. Acetyl-CoA is produced primarily in the mitochondria by β-oxidation of fatty acids or by the action of the pyruvate dehydrogenase complex on pyruvate derived from glycolysis. This mitochondrial acetyl-CoA then enters the TCA cycle to produce citrate, which is subsequently exported from the mitochondria into the cytoplasm and nucleus, where it is converted back into acetyl-CoA for acetylating cytoplasmic and nuclear proteins (Fig. 1) [35, 36]. Thus, the metabolic state of a cell determines the relative levels of acetyl-CoA, a higher level of which might promote increased amounts of protein acetylation as the excess acetyl-CoA not entering the TCA cycle is available to KATs [34].

Given its capability for high-resolution and high-throughput measurements, MS provides a powerful platform for the unbiased analysis of proteins and PTMs. The latter are identified and localized by detecting mass adducts on individual amino acids in peptide fragmentation spectra (MS/MS). Because high-resolution MS is remarkably accurate (<1–2 ppm mass error), it can easily resolve mass shifts caused by isotopes. Thus, it is frequently used in conjunction with labeling techniques based on metabolic incorporation of amino acids or small molecules bearing stable isotopes, such as deuterium, carbon-13, and nitrogen-15. Therefore, an informative way to track the dynamics of PTMs is to exploit the metabolic

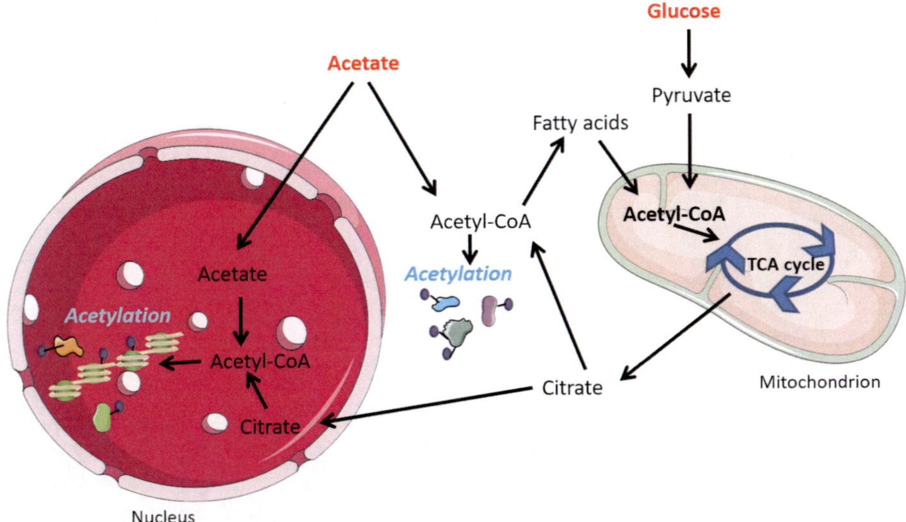

Fig. 1 Metabolic labeling pathway. Acetyl-CoA is produced in the mitochondria from glucose and fatty acids. Mitochondrial acetyl-CoA then enters the tricarboxylic acid (TCA) cycle to produce citrate, which is exported out of the mitochondria into the cytoplasm and nucleus. Citrate is converted back into acetyl-CoA for use in acetylating cytoplasmic and nuclear proteins. Acetate also contributes to both the cytoplasmic and nuclear pool of acetyl-CoA, and is incorporated into acetylation of proteins. The cartoon cell matter were slightly modified from Servier Medical Art [57]

pathways that underlie their synthesis. Under normal cell culture conditions, acetyl-CoA is derived primarily from glucose, though acetate may also contribute in some circumstances [35, 37]. Accordingly, protein acetylation can be labeled using ^{13}C-labeled glucose and acetate, as demonstrated in a study of histone acetylation dynamics [38]. Other PTMs, including methylation and phosphorylation, have also been studied through stable isotope labeling with $^{13}CD_3$-labeled *S*-adenosylmethionine and $^{18}O_4$-labeled ATP, respectively [39, 40]. In contrast to other metabolic labeling strategies, such as SILAC [41], in which heavy amino acids serve as the vehicle by which proteins acquire isotopic labels on the peptide backbone or sidechains but not on PTMs, supplying cells with isotopically labeled small molecule precursors to label PTMs provides additional information. For instance, treating cells with heavy glucose or acetate not only enables an assessment of acetylation dynamics, but also an assessment of which molecule contributes more efficiently to lysine acetylation on a protein-by-protein basis. However, different acetylation sites may exhibit unique dynamics, even when on the same protein, highlighting the necessity of using mass spectrometry to analyze acetylation sites in a broad and high-throughput manner. Mass spectrometry also has the added potential of identifying novel sites of acetylation.

In the following protocol, we describe methods for the quantitative analysis of protein acetylation dynamics (Fig. 2) followed by

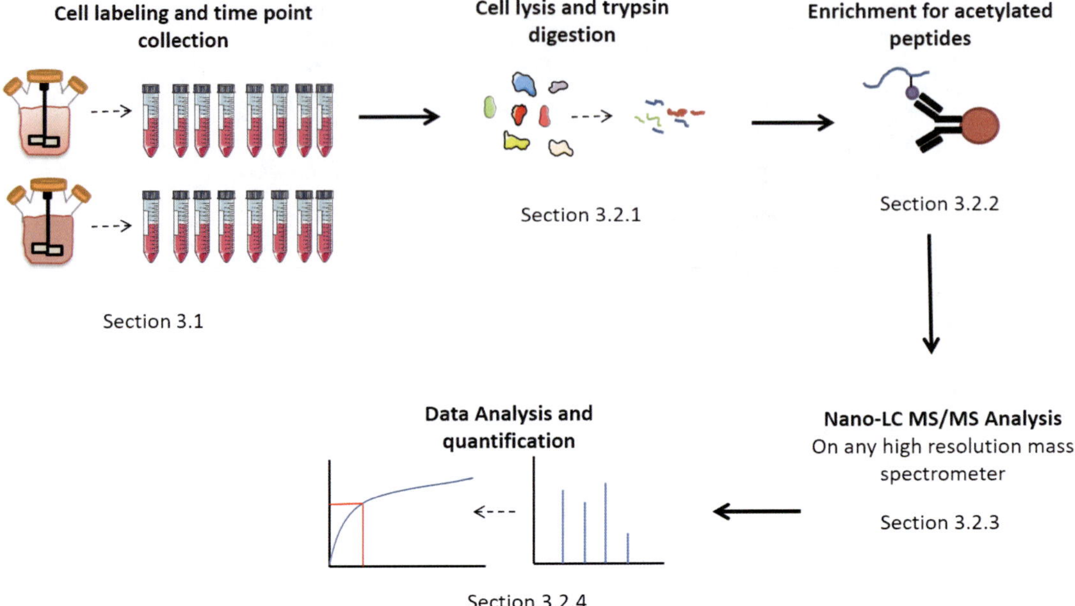

Fig. 2 Workflow for tracking acetylation of non-histone proteins. The basic workflow for studying acetylation dynamics of non-histone proteins involves collection of samples at specified time points, followed by cell lysis and peptide digestion, enrichment of acetylated peptides, and nano-LC MS/MS analysis of the acetylated peptides on any high-resolution mass spectrometer that is suitable for proteomics. The cartoon lab equipment were slightly modified from Servier Medical Art [57]

a focus on histone proteins (Fig. 3) by using MS and stable isotope labeling. Each protocol begins by metabolically labeling cells with stable isotopes (Subheading 3.1). In the case of non-histone proteins, cells are lysed to extract and digest proteins (Subheading 3.2.1) followed by enrichment for acetylated peptides by immunoprecipitation with anti-acetyl antibodies (Subheading 3.2.2) [42]. In the case of histone proteins, cell nuclei are isolated and treated with acid to extract histones, which are then derivatized and digested (Subheading 3.3). Finally, peptides are analyzed by MS to detect (Subheadings 3.2.3 and 3.3.2) and quantify (Subheadings 3.2.4 and 3.3.3) the relative ratio between light and isotope-labeled (heavy) acetyl groups as outlined in Subheading 3.3.3.

2 Materials

2.1 Solutions and Buffers

2.1.1 Solutions and Buffers for General Protein Acetylation Analysis

1. Joklik medium: Joklik Minimum Essential Medium, 10% newborn calf serum (NCS), 1× GlutaMax, 1× Penicillin/Streptomycin.
2. ^{13}C-glucose medium: Joklik Minimum Essential Medium (ingredients list is available online, add everything except light glucose), 2.5 g/L $^{13}C_6$-glucose, 2 mM Alanine, filter using a

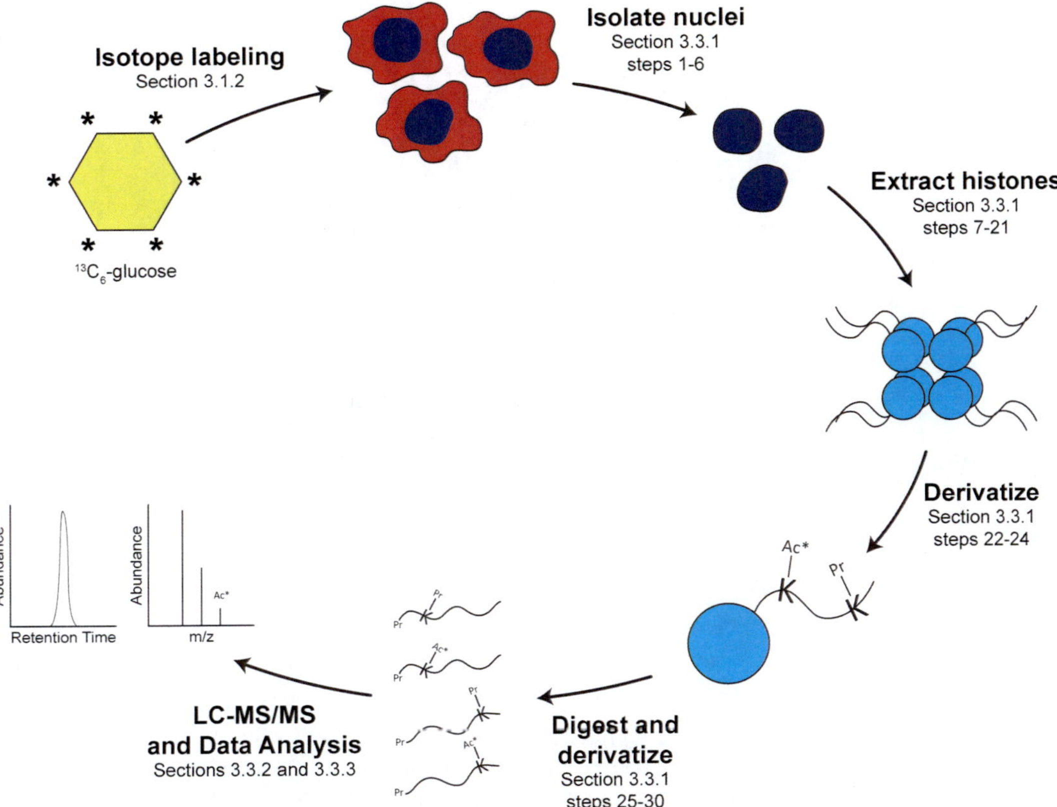

Fig. 3 Workflow for tracking acetylation of histone proteins. To analyze acetylation dynamics on histone proteins, cells are incubated with isotopically labeled glucose (Subheading 3.1.2). After harvesting the cells, nuclei are isolated and then treated with acid to extract histones, which are derivatized with propionic anhydride and digested with trypsin (Subheading 3.3.1). Digested peptides are propionylated once more to derivatize newly generated N-termini and then analyzed by LC-MS/MS (Subheadings 3.3.2 and 3.3.3)

0.2 μm Nalgene disposable filter and then add 10% dialyzed fetal bovine serum (dFBS), 1× GlutaMax, 1× Penicillin/ Streptomycin.

3. D_3-acetate media: Joklik Minimum Essential Medium (ingredients list is available online, add everything including light glucose), 2 mM Alanine, 10 mM D_3-acetate, filter using a 0.2 μm Nalgene disposable filter and then add 10% dialyzed fetal bovine serum (dFBS), 1× GlutaMax, 1× Penicillin/ Streptomycin.

4. Lysis buffer: 6 M urea, 2 M thiourea, 50 mM ammonium bicarbonate, pH 8.2, 1× Halt Protease Inhibitors Cocktail or equivalent, 10 mM sodium butyrate, 10 mM dithiothreitol (DTT).

5. Liquid chromatography coupled to MS (LC-MS) Buffer A: 0.1% formic acid (FA) in MS grade water.
6. LC-MS Buffer B: 80% HPLC-grade acetonitrile (ACN), 0.1% FA in MS grade water.
7. 50 mM triethylammonium bicarbonate (TEAB).
8. Column pre-equilibration buffer for the desalting column procedure: 75% ACN.
9. Equilibration and sample wash buffer for the desalting column procedure: 0.1% Trifluoroacetic acid (TFA).
10. Elution buffer for desalting procedure: 60% ACN, 0.1% TFA.
11. Immunoaffinity purification buffer (IAP) buffer 1×: 50 mM MOPS/NaOH pH 7.2, 10 mM Na_2HPO_4, 50 mM $NaCl^-$.
12. Immunoaffinity purification elution buffer: 0.15% TFA.
13. Stage-tip elution buffer: 75% ACN, 0.1% TFA.
14. 30 mM iodoacetamide (IAA).

2.1.2 Solutions and Buffers for Analysis of Histone Protein Acetylation

1. Growth medium: DMEM with 4.5 g/L glucose, L-glutamine, and sodium pyruvate supplemented with 10% fetal bovine serum and 1× penicillin/streptomycin.
2. Labeling medium: DMEM with L-glutamine without glucose or pyruvate, supplemented with 10% dialyzed fetal bovine serum, 1× penicillin/streptomycin, 2 mM alanine, and desired concentration of unlabeled or labeled substrate, e.g., 25 mM unlabeled glucose or $^{13}C_6$-glucose.
3. Trypsin-EDTA solution (0.25%).
4. Nuclear isolation buffer (NIB): 15 mM Tris pH 7.5, 15 mM NaCl, 60 mM KCl, 5 mM $MgCl_2$, 1 mM $CaCl_2$, 250 mM sucrose, stored at 4 °C and supplemented with 1 mM dithiothreitol (DTT), 500 μM 4-(2-Aminoethyl)benzenesulfonyl fluoride hydrochloride (AEBSF), 5 nM microcystin, and 10 mM sodium butyrate prior to use.
5. 10% NP-40 Alternative.
6. 0.4 N Sulfuric acid H_2SO_4, stored at 4 °C.
7. 100% Trichloroacetic acid (1 g/mL in water), stored at 4 °C.
8. 0.1% HCl in acetone, stored at −20 °C.
9. 0.1 M Ammonium bicarbonate NH_4HCO_3.
10. Trifluoroacetic acid, 1% and 10%.
11. Bradford assay reagent, cuvettes, and spectrophotometer.
12. Reversed-phase (RP) LC buffer A (0.1% formic acid in water).
13. RPLC buffer B (80% acetonitrile with 0.1% formic acid).

2.2 Equipment

2.2.1 Equipment for General Protein Acetylation Analysis

1. Humidified 37 °C incubator with 5% CO_2.
2. 1.5 mL Microcentrifuge tubes.
3. 15 and 50 mL conical tubes.
4. 1 L Spinner flasks.
5. 500 mL Centrifuge bottles.
6. 5 mL Eppendorf tubes.
7. Pipettes (P10, P20, P200, P1000) and tips.
8. −80 °C Freezer.
9. 4 °C Cold room.
10. Centrifuge.
11. Tabletop centrifuge.
12. 0.2 μm Nalgene disposable filter.
13. Cell culture hood.
14. Cell counter.
15. Vortex mixer.
16. Sonicator probe.
17. Spectrophotometer.
18. Rotator.
19. Speed-vac.
20. Sep-Pak C18 Vac columns or equivalent, 50 mg sorbent and 55–105 μm particle size.
21. Vacuum manifold.
22. C_8 filter paper.
23. Repro-Sil Pur C_{18}-AQ 3 μm resin for analytical column for nanoLC.
24. Appropriate nanoflow (nanoLC-MS) setup: HPLC with at least two channels (one for buffer A/loading buffer and one for buffer B) and high-resolution MS.
25. Autosampler vials.

2.2.2 Equipment for Histone Acetylation Analysis

1. Humidified 37 °C incubator with 5% CO_2 atmosphere.
2. Tissue culture flasks and dishes (10 cm).
3. 15 mL Conical tubes.
4. Centrifuge.
5. Liquid nitrogen.
6. Freezer (−80 °C and −20 °C).
7. 1.5 mL Tubes.
8. Rotator.

9. 4 °C Cold room.
10. Pasteur pipettes.
11. C_{18} stage tips and buffers (0.1% trifluoroacetic acid [TFA], 0.5% acetic acid, and 50% acetonitrile with 0.1% formic acid).
12. C_{18} columns (75 μm i.d. × 15–20 cm fused silica packed with 3 μm ReproSil-Pur 120 C_{18}-AQ from Dr. Maisch GmbH).
13. Pressure bomb setup.
14. Appropriate nanoflow (nanoLC-MS) setup: HPLC with at least two channels (one for buffer A/loading buffer and one for buffer B) and high-resolution MS.

3 Methods

3.1 Isotope Labeling of Cell Cultures

3.1.1 Isotope Labeling of Suspension Cell Cultures for General Acetylation Analysis

1. From a single established culture, subculture HeLa suspension cells into four spinner flasks using Joklik medium and grow the cultures until they reach 1 L in volume at 4 × 10^5 cells/mL (*see* **Note 1** for recommendations on number of replicates and information on growing HeLa cells).
2. After reaching 1 L in volume, collect 120 mL of each of the four cultures at a cell confluence of about 4 × 10^5 cells/mL for a baseline unlabeled control (time point 0) for each replicate culture.
3. Pellet cells at 140 × *g* for 5 min in a centrifuge at room temperature. Discard the supernatant.
4. Wash cells by resuspending in 6 mL PBS and centrifuge as in **step 3**. Discard the supernatant.
5. Repeat **step 4** (2 washes total) and then freeze the cell pellet at −80 °C until ready to begin sample processing (Subheading 3.2.1, **step 1**).
6. Pellet the remaining volume of each culture at 140 × *g* for 5 min at room temperature, discard the supernatant, and wash twice with 10 mL PBS. Resuspend two of the four pellets in 1 L of 2.5 g/L ^{13}C-glucose medium. Resuspend the other two pellets in 1 L of 10 mM D_3-acetate medium. In this manner, there are two biological replicates for each labeling method. Note: 2 mM alanine is added to the labeling media to minimize incorporation of $^{13}C_3$-labeled alanine derived from pyruvate [38].
7. Incubate cells at 37 °C. Following the procedures in **steps 3–5**, collect cells from each culture at 0.5, 1, 4, 8, 12, 16, and 24 h after transfer into isotopically labeled medium (*see* **Note 2** for explanation on why specific time points were chosen). Assess cell density and aim to collect approximately 4.8 × 10^7 cells at each time point, which will be around 120 mL for the earlier time points (0–8 h) and 100 mL for the later time points (12–24 h).

3.1.2 Isotope Labeling in Cell Culture for Histone Acetylation Analysis

1. Plate cells in growth medium and incubate at 37 °C overnight such that they will reach roughly 50% confluency the next day (*see* **Note 3**). Prepare one plate for each replicate or treatment. As a control, include at least one plate for untreated cells or, preferably, cells treated with unlabeled substrate (*see* **Note 4**).
2. Remove growth medium and rinse cell monolayer with 10 mL PBS.
3. Add 10 mL of labeling medium containing isotope-labeled substrate and incubate at 37 °C for the desired amount of time (*see* **Note 5**).
4. When ready to harvest cells, transfer the conditioned medium from the cell monolayer to a 15 mL tube, rinse with 10 mL of PBS, add 2 mL of trypsin-EDTA, and swirl briefly to distribute evenly. Incubate for approximately 5 min at room temperature until cells detach. As an alternative to detachment with trypsin, cells may be harvested with a scraper. For cells grown in suspension, collect cells in 15 mL tube and skip to **step 6**.
5. When cells appear rounded and detached, add the conditioned medium (or alternatively, fresh growth medium) back to the 10 cm dish to inactivate the trypsin. Pipette up and down briefly to create a homogenous single cell suspension and transfer to the 15 mL tube.
6. Pellet cells at 1000 rpm (218 × *g*) for 5 min at room temperature.
7. Aspirate the supernatant and gently resuspend the cell pellet in 10 mL of PBS to wash.
8. Pellet cells at 1000 rpm (218 × *g*) for 5 min at room temperature.
9. Aspirate the supernatant.
10. Flash-freeze the cell pellet by carefully immersing in liquid N_2 for 5–10 s.
11. Store cell pellets at −80 °C until ready to proceed with sample preparation.

3.2 Identification of Acetylation Sites and Quantification of Protein Acetylation Rates

3.2.1 Cell Lysis and Trypsin Digestion

1. Lyse the cell pellets in 2.5 times the pellet volume of cold lysis buffer (for example: to a 300 μL pellet add 750 μL lysis buffer) by vortexing followed by sonication on ice in the cold room (*see* **Note 6** for suggestions on sonication).
2. Incubate the samples for 30 min at room temperature.
3. Alkylate reduced cysteine residues with 30 mM iodoacetamide (IAA) for 30 min at room temperature in the dark.
4. A Bradford assay may be conducted to determine protein concentration. Proceed with an equal amount of protein for each time point sample for further processing (*see* **Note 7** for suggestions on protein amount).

5. Digest proteins first with endopeptidase Lys-C for 3 h to enhance digestion efficiency. Use 100 μg (0.38 AU) of Lys-C per 10 mg of protein sample.
6. Dilute the solution with five volumes of 50 mM triethylammonium bicarbonate (TEAB) to dilute out the urea from the lysis buffer, as it will interfere with trypsin digestion.
7. Add sequencing grade trypsin at an enzyme-to-substrate mass ratio of approximately 1:50 and incubate for 12 h at room temperature.
8. Desalt the peptide digest using Sep-Pak C_{18} Vac cartridges, 50 mg sorbent and 55–105 μm particle size and a vacuum manifold. To do so, pre-equilibrate the columns with 3 mL (3× column volume) of 100% ACN and then 2 mL of 75% ACN. Equilibrate with 2 mL 0.1% TFA. Acidify the sample with TFA to pH 3. Load peptide samples and run through column, and then re-load the flow-through on the column for a total of two loads. Wash the loaded sample on the column with 1 mL 0.1% TFA. Elute the sample with 500 μL of 60% ACN + 0.1% TFA into a clean collection tube and repeat once for a total of two elutions.
9. Dry samples using a speed-vac to complete dryness.

3.2.2 Enrichment for Acetylated Peptides

Enrich for acetylated peptides by immunoaffinity purification using the PTMScan Acetyl-Lysine Motif Kit (Cell Signaling Technologies) following the manufacturer's protocol. The protocol is briefly described below.

1. Resuspend the peptide samples in 1× immunoaffinity purification buffer (IAP) (Cell Signaling Technologies) (*see* **Note 8** about saving an input fraction).
2. Wash the antibody-bead slurry four times with 1 mL 1× PBS, centrifuging at 2000 × *g* in a tabletop centrifuge for 30 s to pellet the beads after each wash, and remove previous PBS before adding new PBS for next wash. After the last wash, resuspend the beads as a 50% slurry.
3. Incubate the peptide solutions with the beads overnight at 4 °C while rotating.
4. After incubation, wash the beads with IAP buffer two times, followed by three washes with MS grade water. Complete the wash steps on ice. Between each wash step, centrifuge at 2000 × *g* in a tabletop centrifuge for 30 s to pellet the beads so that the supernatant can be carefully removed.
5. Elute acetylated peptides from the beads using 75 μL of 0.15% TFA, and let stand at room temperature for 10 min, mixing gently by tapping the bottom of the tube several times every

2–3 min. Centrifuge for 30 s at 2000 × *g* in a tabletop centrifuge to collect the elution. Repeat the elution/centrifugation steps for a second elution using 150 μL of 0.15% TFA. After collecting the second elution, combine both eluates in a single tube.

6. Remove any beads that may be present in the eluate by stage-tipping, as described in other studies [43], but with C_8 filter paper to act as a plug. Most of the peptides will flow through the C8 paper while the beads are retained, so make sure to collect the flow-through. Elute any other possible binding peptides from the stage-tips in 75% ACN + 0.1% TFA.
7. Completely dry the elutions in a speed-vac.

3.2.3 NanoLC-MS/MS Analysis of Acetylated Peptides

1. Prepare a picofrit 18 cm long fused silica capillary column (75 μm inner diameter, 360 μm outer diameter) by packing with reversed-phase Repro-Sil Pur C_{18}-AQ 3 μm resin for peptide loading. Preparing a nanoLC column has been described previously [44]. Briefly, cut a 75 μm inner diameter fused silica capillary column at about 30 cm. Pull a tip at one side of the capillary with a laser tip puller. Add Repro-Sil Pur C_{18}-AQ 3 μm resin to a clean HPLC glass vial and resuspend in 100% methanol or another organic solvent for packing the column. Add a miniature stir bar to the glass vial. Place the glass vial with the resin slurry in a pressure (gas) bomb and turn on the magnetic stirrer. Place the pulled silica column in the pressure bomb and turn on the pressure to pack the column. Make sure that the pressure does not exceed the maximum pressure on the gas bomb. After the column is packed, cut the column to a final length of 15–18 cm.
2. Set up a 120 min gradient following the conditions outlined in Table 1. Use a flow rate of 300 nL/min.
3. Set up the MS method to perform data-dependent acquisition (DDA). Here, we describe the method with an Orbitrap Fusion Tribrid mass spectrometer; however, the DDA can be performed with any high-resolution (>20,000 FWHM) MS instrument using an optimized shotgun proteomics method for medium-complex peptide mixtures. If using an Orbitrap Fusion, perform the full MS scan in the Orbitrap at 120,000 resolution (FWHM at 200 *m/z*) with a mass range of 350–1200 *m/z* and an AGC target value of 5×10^5. Perform DDA MS/MS using TopSpeed mode with the Quadrapole as the isolation mode, and the Orbitrap as the detector. Dynamic exclusion should be set for 60 s. Set HCD collision energy to 27, AGC target to 5×10^4, and maximum injection time to 200 ms.

Table 1
Chromatography gradient for analysis of non-histone peptides

Time (min)	Buffer B (%)
0	3
100	38
115	98
120	98

Non-histone peptides may be separated by online nanoLC using the gradient listed

4. Resuspend dried samples in 12 μL of LC-MS buffer A (0.1% formic acid), transfer 11 μL to autosampler vials, and inject 5 μL onto an Easy-nLC system (Thermo) coupled online with an Orbitrap Fusion Tribrid mass spectrometer (Thermo Scientific). Other HPLC and MS systems may be used instead, depending on to which instruments the lab has access.

3.2.4 Database Searching and Quantification of Protein Acetylation

1. Perform database searching with MS raw files using software of choice, such as pFind [45, 46], MaxQuant [47], or vendor associated software. In this specific study we illustrate data processing and analysis by using pFind, setting the search parameters as listed in Table 2.
2. Filter the output file for a list of all of the unique acetylated peptides that were identified along with the precursor *m/z*, charge state, retention time, search score, peptide sequence, modifications, and the respective protein accession numbers.
3. Calculate the fraction of heavy labeled acetylation over total acetylation (heavy acetyl ^{13}C)/(heavy acetyl ^{13}C+ light acetyl ^{12}C) for each acetylated peptide. To calculate the fraction of heavy labeled acetylation over total acetylation, extract the area under the curve (AUC) of the chromatographic peaks corresponding to the monoisotopic peak (M0), the M + 2 peak, and the M + 4 peak (+2 and +4 Da, respectively; *see* Fig. 4 for representative mass spectra). An unlabeled reference for each peptide should be used to calculate the natural abundance of the M + 2 isotope (due to natural occurrence of ^{13}C). This abundance, extracted by integrating the AUC, should be subtracted when estimating the intensity of the labeled acetylation M + 2 peak. In presence of doubly acetylated peptides, this calculation should be repeated also for the M + 4 signal. The sum of M + 2 AUC and M + 4 AUC is divided by the sum of M0, M + 2, M + 4 AUC to calculate ^{13}C/(^{13}C + ^{12}C), the fraction of heavy labeled acetylation.

Table 2
Parameters for database searches using pFind

Parameter	Value
Sequence database	Human UniProt FASTA database [tax id: 9606]
Precursor tolerance for Orbitrap	10 ppm
Fragment tolerance for Orbitrap	0.02 Da
FDR	0.01
Enzyme specificity	Trypsin
Fixed modifications	Cysteine carbamidomethylation (+57.021 Da)
Modifications	Lysine acetylation (+42.026 Da) Methionine oxidation (+15.995 Da)

Peptide identifications may be obtained by running MS raw files through the pFind database search tool using the parameters listed above

4. To visualize the heavy acetylation incorporation, it may be useful to graph the ratio of heavy acetylation incorporation calculated in **step 3** $^{13}C/(^{13}C + ^{12}C)$ plotted against the time as shown in Fig. 5.
5. The half-time of heavy acetyl incorporation can be calculated by fitting the ratios of $^{13}C/(^{13}C + ^{12}C)$ acetylated peptides at each time point to the best fit equation. This can be done using Matlab. Previously, it has been determined by Matlab that the best fit equation is $y = a \times (1-e\wedge(-x/b))$. Here, y and x are the input data: y is the ratio $^{13}C/(^{13}C + ^{12}C)$, and x is the time point (hours). The constants a and b are the output of the fitting curve function, and vary for each peptide based on the data. A Matlab script for calculating the half-life for each peptide given their heavy acetylation ratio at each time point can be distributed upon request.
6. If after the latest time point (i.e., 24 h) the amount of ^{13}C labeled acetylation does not reach 0.5, the linear polynomial equation $y = ax + b$ can be used to calculate the heavy acetylation incorporation half-life.
7. Reproducibility between the two biological replicates across all quantified acetylated peptides can be assessed using Pearson correlation to plot the correlation between time points, and correlation significance (t conversion). We recommend a calculated p-value <0.05 for significance.

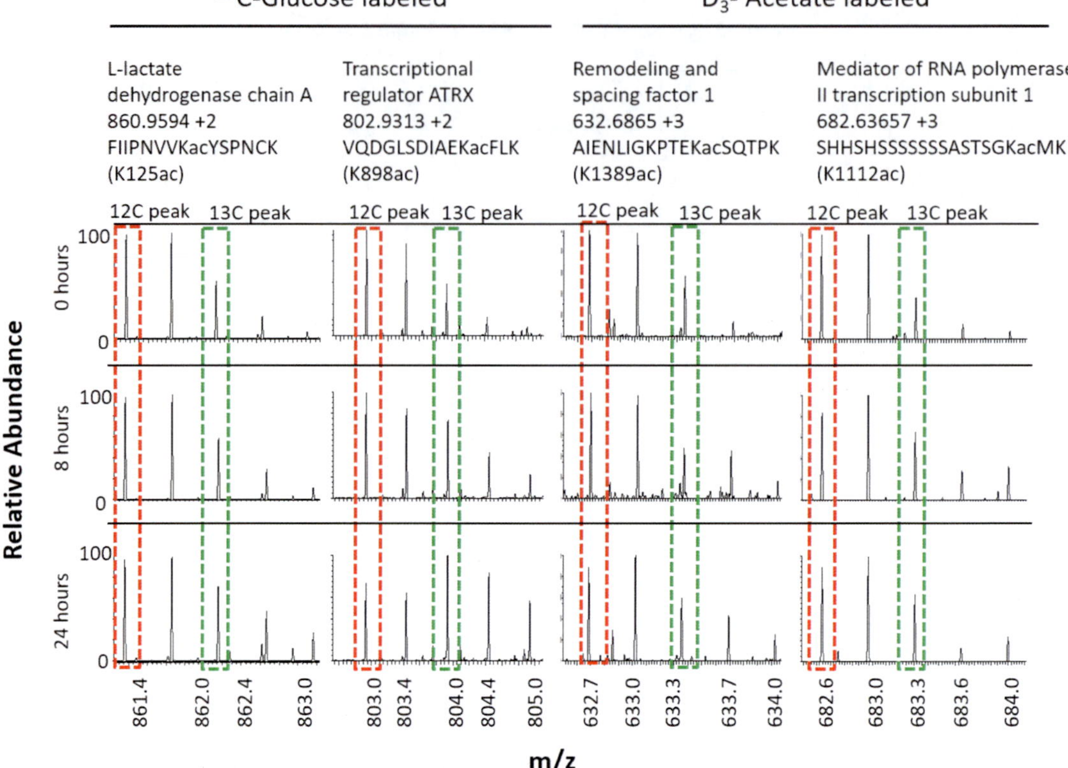

Fig. 4 Representative spectra of acetylated peptides with heavy isotope incorporation from $^{13}C_6$-glucose and D_3-acetate over time. Isotopic patterns indicate heavy acetylation incorporation over time. The red dashed rectangle highlights the monoisotopic peak, and the green dashed rectangle highlights the heavy labeled peak. The isotopic patterns at 0, 8, and 24 h are shown. The L-lactate dehydrogenase peptide by glucose labeling is an example of a peptide that shows minimal incorporation of heavy labeling by 24 h, whereas the ATRX peptide by glucose labeling exemplifies a substantial increase in heavy labeling incorporation by 24 h. The remodeling and spacing factor 1 peptide is an example of a peptide that does not show an increase in heavy labeling incorporation from 0 to 24 h, whereas the mediator of RNA polymerase II transcription subunit 1 peptide does show a mild increase in heavy labeling from 0 to 24 h

8. After converting the quantification values into ratios of heavy form/total heavy + light form, the significance of the monotonic increase of the ratio across the time points can be assessed using the Mann-Kendall test on the average of the two biological replicates (p-value <0.05).
9. Once data are filtered for significantly regulated acetyl sites, further analysis of the data can include, but not be restricted to, hierarchical clustering of trends using, e.g., Perseus [48], Gene Ontology enrichment analysis of acetylated proteins with, e.g., Gorilla [49], visualization of known protein-protein interactions with, e.g., STRING [50], and further network visualizations with, e.g., Cytoscape [51].

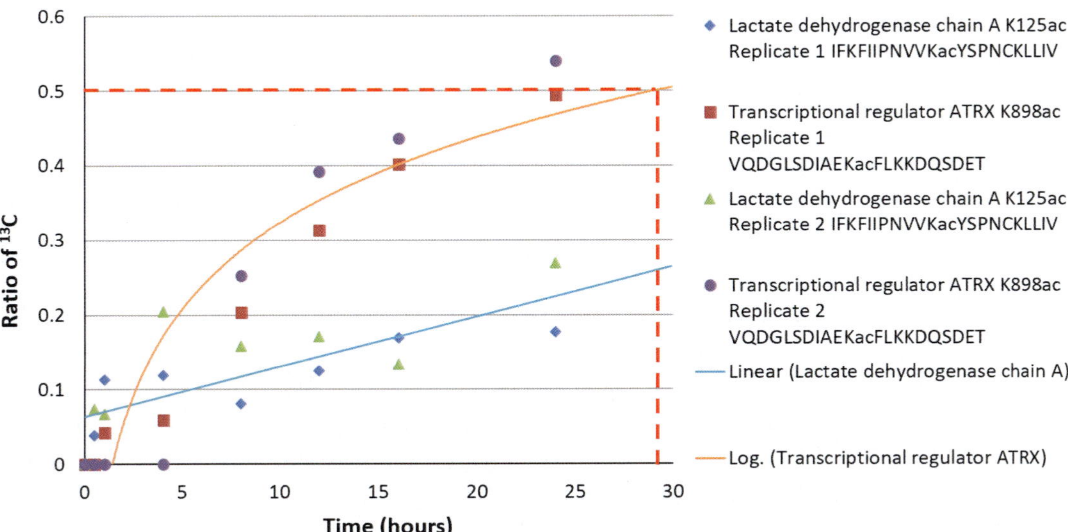

Fig. 5 Analysis of acetylation rates on non-histone proteins. Graph demonstrates the ratio of heavy acetylation incorporation $^{13}C/(^{13}C + ^{12}C)$ plotted against time. The half-time of heavy acetyl incorporation was calculated by fitting the ratios of $^{13}C/(^{13}C + ^{12}C)$ acetylated peptides at each time point to the equation $y = a \times (1-e^\wedge(-x/b))$. If after 24 h the amount of ^{13}C labeled acetylation did not reach 0.5, as shown by the lactate dehydrogenase peptide, we used the linear polynomial equation $y = ax + b$. The red dashed line demonstrates the half-life for the ATRX K898ac site

3.3 Quantitative Analysis of Histone Acetylation Rates

The following procedure is similar to previously published protocols [43, 44, 52, 53].

3.3.1 Histone Extraction, Derivatization, and Proteolytic Digestion

1. Supplement NIB with 1 mM dithiothreitol (DTT), 500 μM 4-(2-Aminoethyl)benzenesulfonyl fluoride hydrochloride (AEBSF), 5 nM microcystin, and 10 mM sodium butyrate prior to use. Prepare 3 mL buffer for every 100 μL of packed cell volume (e.g., 30 mL for ten tubes of 100 μL cells) plus a few milliliter extra to account for pipetting errors. Remove 1/3 of prepared NIB to another tube and to this aliquot, add NP-40 Alternative to a final concentration of 0.2%.
2. Resuspend cell pellet in NIB containing 0.2% NP-40 Alternative. Incubate for 10 min on ice to lyse cells. Use approximately 1 mL buffer per 100 μL of packed cells.
3. Pellet lysed cells at 500 × *g* for 5 min at 4 °C.
4. Remove the supernatant (cytosolic contents) and resuspend the pellet (nuclei) in NIB without NP-40 alternative to wash. Use the same volume of NIB here as used in **step 12**.
5. Pellet nuclei at 500 × *g* for 5 min at 4 °C.
6. Repeat **steps 4** and **5**.

7. Remove the supernatant of the second wash and resuspend the nuclear pellet in 0.4 N (0.2 M) H_2SO_4 to extract histones. Use a volume of H_2SO_4 equal to one-half of the volume of NIB used in **step 1**.
8. Rotate at 4 °C for 2–4 h.
9. Pellet insoluble debris by centrifugation at 3400 × *g* for 5 min at 4 °C.
10. Transfer the supernatant, which contains histones, to a 1.5 mL tube. Precipitate proteins by adding one volume of 100% trichloroacetic acid to three volumes of supernatant (e.g., 167 μL trichloroacetic acid to 500 μL acid extract).
11. Mix well and incubate overnight on ice in a 4 °C cold room.
12. Pellet precipitated protein by centrifugation at 3400 × *g* for 5 min at 4 °C.
13. Remove supernatant and wash pellet by adding approximately 500 μL cold 0.1% HCl in acetone with a glass Pasteur pipette.
14. Centrifuge as in **step 12**.
15. Remove supernatant and wash pellet by adding approximately 500 μL cold acetone with a glass Pasteur pipette.
16. Centrifuge as in **step 12**.
17. Remove supernatant, leave the tube uncapped, and allow the pellet to air-dry for 5–10 min on the bench.
18. Resuspend the pellet in 0.1 M NH_4HCO_3. Use approximately 20–40 μL per 100 μL of original cell pellet, which should result in a protein concentration of roughly 2–4 μg/μL. Check pH with pH paper and, if necessary, add additional NH_4HCO_3 salt or concentrated (e.g., 1 M) NH_4HCO_3 so that the pH is 7–8.
19. Pellet insoluble debris by centrifugation at 17,000 × *g* for 5 min at room temperature.
20. Measure protein concentration of supernatant by Bradford assay.
21. For each sample, place 20 μg of protein in a 1.5 mL tube. Add 0.1 M NH_4HCO_3 to a final volume of 20 μL. *See* **Note 9**.
22. Add 10 μL of 25% propionic anhydride in 2-propanol (e.g., 10 μL propionic anhydride to 30 μL 2-propanol) to each 20 μL sample (scale accordingly if using different volumes). Add 2–3 scoops of NH_4HCO_3 salt using a beveled P1000 pipette tip to counteract acidification caused by the reaction as the derivatization is less efficient at acidic pH. Ammonium hydroxide may also be used [43, 44, 52, 53]. The pH of the reaction should be approximately 8. Incubate for 15 min in a 37 °C water bath. *See* **Note 10**.
23. Dry samples in a speed-vac. *See* **Note 11**.
24. Resuspend samples in 20 μL of 0.1 M NH_4HCO_3 and repeat **steps 22** and **23**.

25. Resuspend samples in 20 μL of 0.1 M NH_4HCO_3, ensure the pH is 7–8 with pH paper, and add 1 μg trypsin for every 20 μg of protein.
26. Incubate overnight at room temperature.
27. Repeat **steps 22** and **23** twice to derivatize the newly generated N-termini by propionylation.
28. Resuspend samples in 1% trifluoroacetic acid (TFA) and add additional concentrated 10% TFA as necessary so that the pH is 2–3. Pellet insoluble debris by centrifugation at 17,000 × *g* for 5 min at room temperature.
29. Desalt supernatant with C_{18} stage tips. *See* **Note 12**.
30. Dry samples in speed-vac and resuspend in 0.1% formic acid at approximately 1 μg/μL. Pellet insoluble debris by centrifugation at 17,000 × *g* for 5 min at room temperature, and then transfer the supernatant to an autosampler vial and proceed to analysis by LC-MS/MS. *See* **Note 13**.

3.3.2 Nano LC-MS/MS Analysis of Acetylated Histone Peptides

1. Pack a column with C_{18} material and connect it to a nano liquid chromatography system coupled to a mass spectrometer.
2. Set the chromatography gradient as listed in Table 3. Use a flow rate of 300 nL/min.
3. Program the mass spectrometer for data-independent acquisition (DIA), which results in fragmentation of all peptide ions at a given retention time to facilitate quantitation of isobaric peptides bearing the same marks at different positions. To cover a mass range from 300 to 1100 m/z, each instrument cycle consists of 16 DIA MS/MS scans of 50 m/z isolation width with two full MS scans, one at the beginning and one after the eighth MS/MS scan [53, 54]. Full scans are performed with the orbitrap in positive profile mode from 300 to 1100 m/z at a resolution of 120,000. DIA MS/MS scans are performed in the ion trap in centroid mode using CID

Table 3
Chromatography gradient for analysis of histone peptides

Time (min)	Buffer B (%)
0	5
45	33
50	98
60	98

Histone peptides may be separated by online nanoLC using the gradient listed above

Table 4
Mass spectrometry scan settings for the analysis of histone peptides by data-independent acquisition (DIA)

Scan #	Type	Isolation window center (*m/z*)	Mass range (*m/z*)
1	Full MS	–	300–1100
2	DIA MS/MS	325	120–1500
3	DIA MS/MS	375	120–1500
4	DIA MS/MS	425	120–1500
5	DIA MS/MS	475	130–1500
6	DIA MS/MS	525	140–1500
7	DIA MS/MS	575	155–1500
8	DIA MS/MS	625	170–1500
9	DIA MS/MS	675	185–1500
10	Full MS	–	300–1100
11	DIA MS/MS	725	195–1500
12	DIA MS/MS	775	210–1500
13	DIA MS/MS	825	225–1500
14	DIA MS/MS	875	240–1500
15	DIA MS/MS	925	250–1500
16	DIA MS/MS	975	265–1500
17	DIA MS/MS	1025	280–1500
18	DIA MS/MS	1075	295–1500

Histone peptides may be analyzed by tandem mass spectrometry using the instrument cycle shown

fragmentation with a normalized collision energy of 35%, activation Q of 0.25, activation time of 10 ms, and maximum injection time of 50 ms. The mass range of each MS/MS scan is varied slightly depending on the isolation window. Table 4 outlines the scan settings.

4. Inject approximately 1 μg of histone peptides and collect data. *See* **Note 14**.

3.3.3 MS Data Analysis for Quantification of Histone Acetylation Analysis

1. We analyze MS data with a custom Matlab script called EpiProfile [55], which reports the chromatographic peak areas of acetylated histone peptides bearing either unlabeled or $^{13}C_2$-labeled acetyl groups and then calculates the relative abun-

dance of each. For a given acetylation event (e.g., H4K12ac), the percent of ^{13}C-labeling is obtained by dividing the relative abundance of the ^{13}C-labeled form by the sum of the relative abundances of the ^{13}C-labeled and unlabeled forms. The script is freely available from the Garcia Lab upon request (*see* http://hosting.med.upenn.edu/garcialab/resources/). Any labeling due to naturally occurring ^{13}C can be subtracted out based on the percent ^{13}C-labeled peptide found in the unlabeled control cells. Peak areas may also be manually extracted, using software such as Xcalibur, based on the expected m/z value of a peptide of interest, keeping in mind to add the mass of a propionyl group (+56.0262 Da) for the N-terminus and every unmodified or mono-methylated lysine residue, the mass of a methyl group (+14.0157 Da) for every methylation event (e.g., multiply by 3 for tri-methylation), the mass of an acetyl group for every acetylation event (+42.0106), and the mass of a proton (+1.0073) based on the charge state z (usually +2 but occasionally +3). The ^{13}C-labeled form of an acetylated peptide should be $(+2n)/z$ units (Daltons) greater than the unlabeled form, where n equals the number of labeled acetyl groups. As before, the relative abundance of isotopic labeling is calculated by dividing the peak area of the ^{13}C-labeled peptide by the sum of the peak areas of the unlabeled and ^{13}C-labeled forms. *See* **Notes 15** and **16**. Representative mass spectra data are shown in Fig. 6. An example of a plot showing acetylation kinetics for peptides from H3 is depicted in Fig. 7.

4 Notes

1. At least two replicates are strongly suggested for each isotopically labeled substrate, which is why this protocol calls for splitting the HeLa cultures into four separate cultures. However, the number of replicates can be varied by splitting the HeLa cultures accordingly. The optimal concentration for HeLa cell growth is 2×10^5 cells/mL. Culture the cells by diluting them to 2×10^5 cells/mL with Joklik medium every day. We recommend starting this labeling experiment at 4×10^5 cells/mL so that there will be a larger yield of protein for each time point without the cells becoming overly dense at the later time points. Variation between 2 and 4×10^5 cells/mL is fine, but avoid starting with a culture that is denser than 4×10^5 cells/mL because the culture will be overgrown by the later time points, which may affect cell viability. Growing cells to a volume of 1 L will allow 120 mL of cells (at an approximate density of 4×10^5 cells/mL for a total of 4.8×10^7) to be collected across 8 time points. If larger amounts of starting material are desired, it is possible to grow the HeLa cultures to a volume

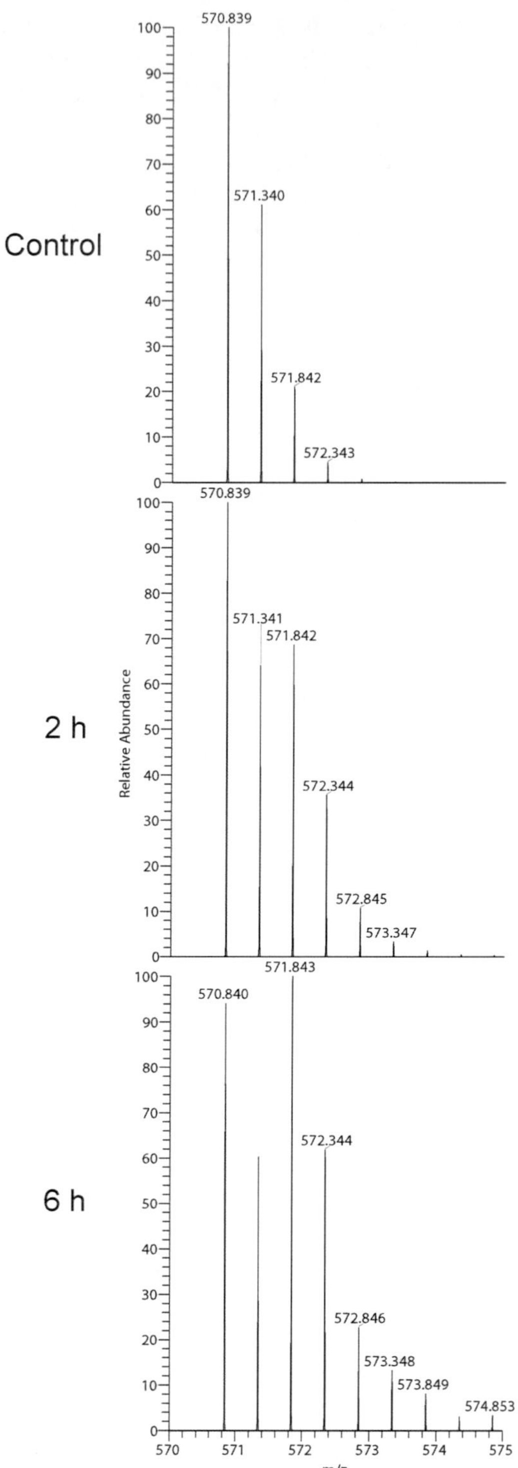

Fig. 6 Representative spectra of acetylated histone peptides. The mass spectrum of the doubly charged H3 peptide ^{18}KQLATKAAR^{26} bearing acetylation at K18 is shown for HEK293 cells treated with unlabeled glucose (Control) or with 25 mM $^{13}C_6$-glucose for various amounts of time. The relative abundance of the +2 Da peak at 571.84 *m/z* increases over time, indicating the presence of an acetyl group with two carbon-13 atoms

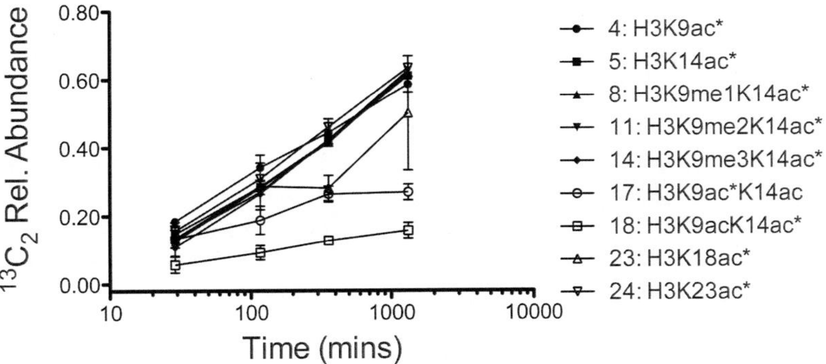

Fig. 7 Kinetic analysis of histone acetylation. The relative abundance of the isotope-labeled species is plotted over time for several distinctly modified H3 peptides from HEK293 cells treated with 25 mM $^{13}C_6$-glucose (mean ± standard deviation of technical replicates). The residue bearing the labeled acetyl group is indicated in the legend with an asterisk

larger than 1 L so that a larger volume may be collected at each time point.

2. The time points 0.5, 1, 4, 8, 12, 16, and 24 h, as listed in this study, are recommended in order to get a better analysis of heavy acetylation incorporation trends. The 0 time point is the control sample that is harvested from the unlabeled media, as described in the protocol. The specific time points chosen and the number of time points may be varied, but it is recommended that at least 6 time points are used, as 6 degrees of freedom has good statistical power in correlation analysis. Additionally, the initial time points after switching to heavy media should be closer together (0, 0.5, 1, 4) compared to the later time points in order to get a better record of the initial incorporation. Time points up to 24 h are recommended because it has been previously shown in HeLa cells that the average turnover rate of proteins is ~20 h [56].
3. For instance, add 3 × 10^6 HEK293 cells to 10 mL of growth medium in a 10 cm dish. Scale appropriately for different sizes of culture vessels. This number of HEK293 cells should yield approximately 50–100 μg of histone extract, which is well above the target of 20 μg.
4. To monitor kinetics, a time course may be performed by adding substrate and then harvesting cells at various times after addition. Alternatively, a reverse time course may be performed in which substrate is added at various times ahead of a set time for harvesting all treatments at once (e.g., add substrate at t – 24 h and t – 2 h and harvest at time t).
5. The substrate and its concentration are specified by the user based on the desire to know how much a given carbon source

contributes to acetylation. The only requirement is that the substrate provides carbon for acetyl-CoA production. We have specified $^{13}C_6$-labeled glucose as an example as mentioned in Subheading 2.1.2 in the labeling medium composition. In the case of isotopic labeling with $^{13}C_6$-glucose, exogenous unlabeled alanine may be added to suppress the appearance of the isotope label on amino acids instead of the acetyl group due to conversion of glucose to alanine via pyruvate [38].

6. Sonication of cells in lysis buffer was performed at power setting 4 on a Sonic Dismembrator Model 100 for 3 cycles of 10 s with 30 s breaks in between. The duration or number of sonication cycles may be varied as necessary to reduce sample viscosity. Remember to keep the tubes in ice in the cold room, as sonication generates heating, which could result in chemical derivatization of the proteins by urea.
7. For large-scale analysis, a total amount of 10 mg of protein per sample is suggested to use as input for the digestion. However, the amount of protein used for digestion can be varied, depending on the amount of total protein extracted from the cells at each time point. It is important that an equal amount of protein is used for each time point for digestion and further processing.
8. If you would like to calculate the percent enrichment achieved using the anti-acetyl lysine antibodies, save a fraction of the input from each of your time point samples before adding them to the anti-acetyl lysine beads. Analyze these input fractions by MS (with the same MS conditions as for the time point samples) in order to compare the number of acetylated peptide identifications in the enriched sample versus the input sample. In addition, saving an input fraction is useful to account for any changes in protein expression levels that may appear as changes in acetylation levels.
9. Alternative amounts and volumes may be used, though starting with less than 10–20 μg may impact the quality of results. Excessively large volumes will also require additional drying time. Any remaining histones may be stored at −80 °C.
10. In this step, N-termini and unmodified and mono-methylated lysine residues are derivatized by propionylation to block trypsin cleavage at lysine residues, which, given the high lysine content of histones, would otherwise produce smaller peptides that are not suitable for LC-MS/MS analysis. Additionally, the propionyl group increases peptide hydrophobicity, resulting in better retention by the C_{18} resin. Prepare fresh propionylation reagent every three to five samples and complete this step rapidly by performing the whole process (up to the incu-

bation) on no more than three to five samples at a time. Be aware that adding NH_4HCO_3 salt to the reaction will cause some bubbling as gas is released, which could lead to a build-up of pressure in the tube and make the cap prone to popping, though this rarely happens in our experience. To extend its shelf life, propionic anhydride can be stored under a thin layer of argon gas.

11. Drying times may vary greatly, and sometimes, a translucent pellet remains. In our experience, adding more ammonium bicarbonate salt seems to result in faster drying times.

12. Briefly, we create C_{18} stage tips by cutting out a circular piece (1–2 mm diameter) of C_{18} material (C_{18} Empore, 3 M) with a shortened P1000 pipette tip (cut a few mm off the end with a razor blade to obtain a larger diameter bore) and packing it into a P200 pipette tip with a piece of capillary. The stage tip is mounted with an adaptor on 1.5 mL tubes for centrifugation at 1000 × *g* for 1 min (or more time or higher speed as necessary) to force liquid through the resin. The resin is activated with 30 μL methanol, equilibrated twice with 30 μL 0.1% TFA, loaded with sample, and washed twice with 30 μL 0.1% TFA and once with 30 μL of 0.5% acetic acid with the collected liquid being discarded as necessary such that the tip does not contact the waste liquid. The desalted peptides are then eluted with 30 μL 50% ACN/0.1% formic acid into a clean 1.5 mL tube.

13. We assume 50% sample loss throughout the procedure. We typically resuspend the dried peptides in 12 μL, pellet the insoluble debris, and then remove 10 μL to the sample vial to avoid transferring any particulates into the sample vial, which could clog the LC column. We also centrifuge the sample vials briefly (1–2 min at 3400 × *g*) before placing in the autosampler for the same reason and to eliminate any air bubbles. If peptides will not be analyzed immediately, they can be stored at −20 °C.

14. We typically inject 2 μL of sample. If the protein yield was low in **step 31**, a larger volume can be injected.

15. Beyond the incorporation of $^{13}C_2$-labeled acetyl groups, an additional cause of a shift in the isotopic distribution, which is not accounted for here, may be the incorporation of $^{13}C_3$-labeled alanine derived from pyruvate [38]. Alanine is added to the labeling medium to minimize the contribution of this pathway. Fragmentation spectra can be inspected to confirm that the isotopic shift is due to labeling of acetylated lysines.

16. Statistical analysis of ^{13}C isotope incorporation into histone peptides may be performed in the same manner as in the case of non-histone proteins (Subheading 3.2.4).

Acknowledgments

Funding support from NIH grants 2T32CA009140-41A1, R01GM110174, R01AI118891, P01CA196539, and T32 GM071399 is gratefully acknowledged.

References

1. Kim G-W, Yang X-J (2010) Comprehensive lysine acetylomes emerging from bacteria to humans. Trends Biochem Sci 36:211–220
2. DesJarlais R, Tummino PJ (2016) Role of histone-modifying enzymes and their complexes in regulation of chromatin biology. Biochemistry 55:1584–1599
3. Phillips DM (1963) The presence of acetyl groups of histones. Biochem J 87:258–263
4. Allfrey VG, Faulkner R, Mirsky AE (1964) Acetylation and methylation of histones and their possible role in the regulation of RNA synthesis. Proc Natl Acad Sci U S A 51:786–794
5. Luger K, Mäder AW, Richmond RK et al (1997) Crystal structure of the nucleosome core particle at 2.8 A resolution. Nature 389:251–260
6. McGinty RK, Tan S (2014) Nucleosome structure and function. Chem Rev 115:2255–2273
7. Berger SL (2007) The complex language of chromatin regulation during transcription. Nature 447:407–412
8. Kouzarides T (2007) Chromatin modifications and their function. Cell 128:693–705
9. Bannister AJ, Kouzarides T (2011) Regulation of chromatin by histone modifications. Cell Res 21:381–395
10. Kebede AF, Schneider R, Daujat S (2014) Novel types and sites of histone modifications emerge as players in the transcriptional regulation contest. FEBS J 282:1658–1674
11. Zhao Y, Garcia BA (2015) Comprehensive catalog of currently documented histone modifications. Cold Spring Harb Perspect Biol 7:a025064
12. Strahl BD, Allis CD (2000) The language of covalent histone modifications. Nature 403:41–45
13. Falkenberg KJ, Johnstone RW (2014) Histone deacetylases and their inhibitors in cancer, neurological diseases and immune disorders. Nat Rev Drug Discov 13:673–691
14. Dhalluin C, Carlson JE, Zeng L et al (1999) Structure and ligand of a histone acetyltransferase bromodomain. Nature 399:491–496
15. Sanchez R, Meslamani J, Zhou M-M (2014) The bromodomain: from epigenome reader to druggable target. Biochim Biophys Acta 1839:676–685
16. Tessarz P, Kouzarides T (2014) Histone core modifications regulating nucleosome structure and dynamics. Nat Rev Mol Cell Biol 15:703–708
17. Glozak MA, Sengupta N, Zhang X et al (2005) Acetylation and deacetylation of non-histone proteins. Gene 363:15–23
18. Kim SC, Sprung R, Chen Y et al (2006) Substrate and functional diversity of lysine acetylation revealed by a proteomics survey. Mol Cell 23:607–618
19. Choudhary C, Kumar C, Gnad F et al (2009) Lysine acetylation targets protein complexes and co-regulates major cellular functions. Science 325:834–840
20. Ito A, Lai CH, Zhao X et al (2001) p300/CBP-mediated p53 acetylation is commonly induced by p53-activating agents and inhibited by MDM2. EMBO J 20:1331–1340
21. Rausa FM, Hughes DE, Costa RH (2004) Stability of the hepatocyte nuclear factor 6 transcription factor requires acetylation by the CREB-binding protein coactivator. J Biol Chem 279:43070–43076
22. Li K, Wang R, Lozada E et al (2010) Acetylation of WRN protein regulates its stability by inhibiting ubiquitination. PLoS One 5:e10341
23. Hernandez-Hernandez A, Ray P, Litos G et al (2006) Acetylation and MAPK phosphorylation cooperate to regulate the degradation of active GATA-1. EMBO J 25:3264–3274
24. Spange S, Wagner T, Heinzel T et al (2008) Acetylation of non-histone proteins modulates cellular signalling at multiple levels. Int J Biochem Cell Biol 41:185–198
25. Blanco-García N, Asensio-Juan E, de la Cruz X et al (2008) Autoacetylation regulates P/CAF nuclear localization. J Biol Chem 284: 1343–1352
26. Thevenet L, Méjean C, Moniot B et al (2004) Regulation of human SRY subcellular distribu-

tion by its acetylation/deacetylation. EMBO J 23:3336–3345
27. Svejstrup JQ (2007) Elongator complex: how many roles does it play? Curr Opin Cell Biol 19:331–336
28. Simpson CL, Lemmens R, Miskiewicz K et al (2008) Variants of the elongator protein 3 (ELP3) gene are associated with motor neuron degeneration. Hum Mol Genet 18:472–481
29. Narayan S, Bader GD, Reimand J (2016) Frequent mutations in acetylation and ubiquitination sites suggest novel driver mechanisms of cancer. Genome Med 8:55
30. Mertins P, Qiao JW, Patel J et al (2013) Integrated proteomic analysis of post-translational modifications by serial enrichment. Nat Methods 10:634–637
31. Fan J, Krautkramer KA, Feldman JL et al (2015) Metabolic regulation of histone post-translational modifications. ACS Chem Biol 10:95–108
32. Lu C, Thompson CB (2012) Metabolic regulation of epigenetics. Cell Metab 16:9–17
33. Janke R, Dodson AE, Rine J (2015) Metabolism and epigenetics. Annu Rev Cell Dev Biol 31:473–496
34. Donohoe DR, Collins LB, Wali A et al (2012) The Warburg effect dictates the mechanism of butyrate-mediated histone acetylation and cell proliferation. Mol Cell 48:612–626
35. Wellen KE, Hatzivassiliou G, Sachdeva UM et al (2009) ATP-citrate lyase links cellular metabolism to histone acetylation. Science 324:1076–1080
36. Choudhary C, Weinert BT, Nishida Y et al (2014) The growing landscape of lysine acetylation links metabolism and cell signalling. Nat Rev Mol Cell Biol 15:536–550
37. Kamphorst JJ, Chung MK, Fan J et al (2014) Quantitative analysis of acetyl-CoA production in hypoxic cancer cells reveals substantial contribution from acetate. Cancer Metab 2:23
38. Evertts AG, Zee BM, Dimaggio PA et al (2013) Quantitative dynamics of the link between cellular metabolism and histone acetylation. J Biol Chem 288:12142–12151
39. Zee BM, Levin RS, Xu B et al (2009) In vivo residue-specific histone methylation dynamics. J Biol Chem 285:3341–3350
40. Molden RC, Goya J, Khan Z et al (2014) Stable isotope labeling of phosphoproteins for large-scale phosphorylation rate determination. Mol Cell Proteomics 13:1106–1118
41. Ong S-E, Blagoev B, Kratchmarova I et al (2002) Stable isotope labeling by amino acids in cell culture, SILAC, as a simple and accurate approach to expression proteomics. Mol Cell Proteomics 1:376–386
42. Svinkina T, Gu H, Silva JC et al (2015) Deep, quantitative coverage of the lysine acetylome using novel anti-acetyl-lysine antibodies and an optimized proteomic workflow. Mol Cell Proteomics 14:2429–2440
43. Sidoli S, Bhanu NV, Karch KR et al (2016) Complete workflow for analysis of histone post-translational modifications using bottom-up mass spectrometry: from histone extraction to data analysis. J Vis Exp 111:54112
44. Sidoli S, Garcia BA (2016) Characterization of individual histone posttranslational modifications and their combinatorial patterns by mass spectrometry-based proteomics strategies. Methods Mol Biol 1528:121–148
45. Chi H, He K, Yang B et al (2015) pFind-Alioth: a novel unrestricted database search algorithm to improve the interpretation of high-resolution MS/MS data. J Proteome 125:89–97
46. Wang L-H, Li D-Q, Fu Y et al (2007) pFind 2.0: a software package for peptide and protein identification via tandem mass spectrometry. Rapid Commun Mass Spectrom 21:2985–2991
47. Cox J, Mann M (2008) MaxQuant enables high peptide identification rates, individualized p.p.b.-range mass accuracies and proteome-wide protein quantification. Nat Biotechnol 26:1367–1372
48. Tyanova S, Temu T, Sinitcyn P et al (2016) The Perseus computational platform for comprehensive analysis of (prote)omics data. Nat Methods 13:731–740
49. Eden E, Navon R, Steinfeld I et al (2009) GOrilla: a tool for discovery and visualization of enriched GO terms in ranked gene lists. BMC Bioinformatics 10:48
50. Szklarczyk D, Franceschini A, Wyder S et al (2014) STRING v10: protein-protein interaction networks, integrated over the tree of life. Nucleic Acids Res 43:D447–D452
51. Shannon P, Markiel A, Ozier O et al (2003) Cytoscape: a software environment for integrated models of biomolecular interaction networks. Genome Res 13:2498–2504
52. Lin S, Garcia BA (2012) Examining histone posttranslational modification patterns by high-resolution mass spectrometry. Methods Enzymol 512:3–28
53. Karch KR, Sidoli S, Garcia BA (2016) Identification and quantification of histone

PTMs using high-resolution mass spectrometry. Methods Enzymol 574:3–29

54. Sidoli S, Simithy J, Karch KR et al (2015) Low resolution data-independent acquisition in an LTQ-Orbitrap allows for simplified and fully untargeted analysis of histone modifications. Anal Chem 87:11448–11454
55. Yuan Z-F, Lin S, Molden RC et al (2015) EpiProfile quantifies histone peptides with modifications by extracting retention time and intensity in high-resolution mass spectra. Mol Cell Proteomics 14:1696–1707
56. Boisvert F-M, Ahmad Y, Gierliński M et al (2011) A quantitative spatial proteomics analysis of proteome turnover in human cells. Mol Cell Proteomics 11:M111.011429
57. Servier (2017) Powerpoint image bank. http://www.servier.com/Powerpoint-image-bank. License. https://creativecommons.org/licenses/by/3.0/legalcode

Chapter 6

Quantitative Analysis of Protein *S*-Acylation Site Dynamics Using Site-Specific Acyl-Biotin Exchange (ssABE)

Keith T. Woodley and Mark O. Collins

Abstract

Protein *S*-acylation (palmitoylation) is a reversible lipid modification that is increasingly recognized as an important regulator of protein function, including membrane association, trafficking, and subcellular localization. Most proteomic methods to study palmitoylation allow characterization of putative palmitoylated proteins but do not permit identification of individual sites of palmitoylation. We have recently adapted the Acyl-Biotin Exchange (ABE) method that is routinely used for palmitoyl-proteome characterization, to permit global *S*-acylation site analysis. This site-specific ABE (ssABE) protocol, when combined with SILAC-based quantification, allows both the large-scale identification of palmitoylation sites and quantitative profiling of palmitoylation site changes. This approach enables palmitoylation to be studied at a systems level comparable to other more intensively studied post-translational modifications.

Key words Protein, *S*-Acylation, Palmitoylation, Membrane, Proteomic, Mass spectrometry, Quantitative

1 Introduction

Protein *S*-acylation, also known as palmitoylation, is the attachment of a fatty acid to the side chain of a cysteine residue through a thioester bond. It is an important post-translational modification within the cell as it is the only reversible lipid modification and is mainly used as a lipid anchor for membrane attachment [1, 2]. The ability to identify *S*-acylated proteins on a proteome-wide scale has led to this modification being recognized as widespread and involved in a number of processes, with up to 10% of genes having a protein product that is *S*-acylated [3]. There are a number of methods to identify *S*-acylated proteins which can be divided into three types: removal of the fatty acid and replacement with an affinity tag (often biotin), known as acyl-biotin exchange (ABE) [4], removal of the fatty acid and capture on a thiol-reactive resin (acyl-RAC) [5] or metabolic labeling of *S*-acylation sites with a clickable fatty acid analogue and click chemistry to replace the fatty

Caroline A. Evans et al. (eds.), *Mass Spectrometry of Proteins: Methods and Protocols*, Methods in Molecular Biology, vol. 1977,
https://doi.org/10.1007/978-1-4939-9232-4_6, © Springer Science+Business Media, LLC, part of Springer Nature 2019

acid analogue with an affinity tag, known as metabolic labeling-click chemistry (MLCC) [6]. Each method has been shown to enrich an overlapping but distinct subset of the palmitoylated proteome [7], but ABE has the advantage that it can be performed on tissue samples as it does not require metabolic and the coverage of palmitoyl-proteomes is higher than that obtained using metabolic labeling [7].

One of the major drawbacks of the classical ABE method is that it only provides information about potential *S*-acylation on a protein level. We have developed an ABE-based method which allows site-specific analysis of *S*-acylation, allowing the identification of a large number of *S*-acylation sites from a single LC-MS/MS run [8]. Using this method a significant number of known sites were identified, as well as hundreds of novel sites on known *S*-acylated proteins, allowing for the first time global analysis of *S*-acylation site on a proteome scale. This method can be used to probe protein *S*-acylation in any organism, tissue, or cell type and can be combined with a number of quantification types including SILAC (Stable isotope labeling with amino acids in cell culture) to monitor *S*-acylation site dynamics.

The protocol described here employs SILAC-based quantitation which affords high accuracy quantification that can measure small changes in palmitoylation, which is particularly important in perturbation experiments. ssABE may be performed using label-free quantitation [8] to identify putative sites of palmitoylation and in perturbation experiments where measurement of small changes (less than twofold) is not required. A schematic of the workflow for ssABE with SILAC-based quantitation is shown in Fig. 1. The workflow for ssABE with label-free quantification is almost identical to ssABE with SILAC-based quantitation, with the exception that samples are not pooled prior to LC-MS/MS analysis; the control and hydroxylamine-treated samples are analyzed in two separate LC-MS/MS acquisitions. In perturbation experiments, samples to be compared may be differentially SILAC labeled, pooled in **step 1** of the protocol, thereby minimizing experimental variation (*see* Subheading 3.8). A detailed description of SILAC-based quantification is not included in this protocol as it is the subject of a dedicated volume of Methods in Molecular Biology entitled “Stable Isotope Labelling by Amino Acids in Cell Culture (SILAC)”.

2 Materials

2.1 Protein Extraction, Reduction, and Alkylation

1. Extraction buffer: 4% SDS, 100 mM Tris pH 8.5, 2 μg/μL aprotinin/leupeptin, 0.5 mM PMSF (*see* **Note 1**), 20 μM ZnCl, 5 mM EDTA, 50 mM TCEP, and 25 mM iodoacetamide (prepared fresh).

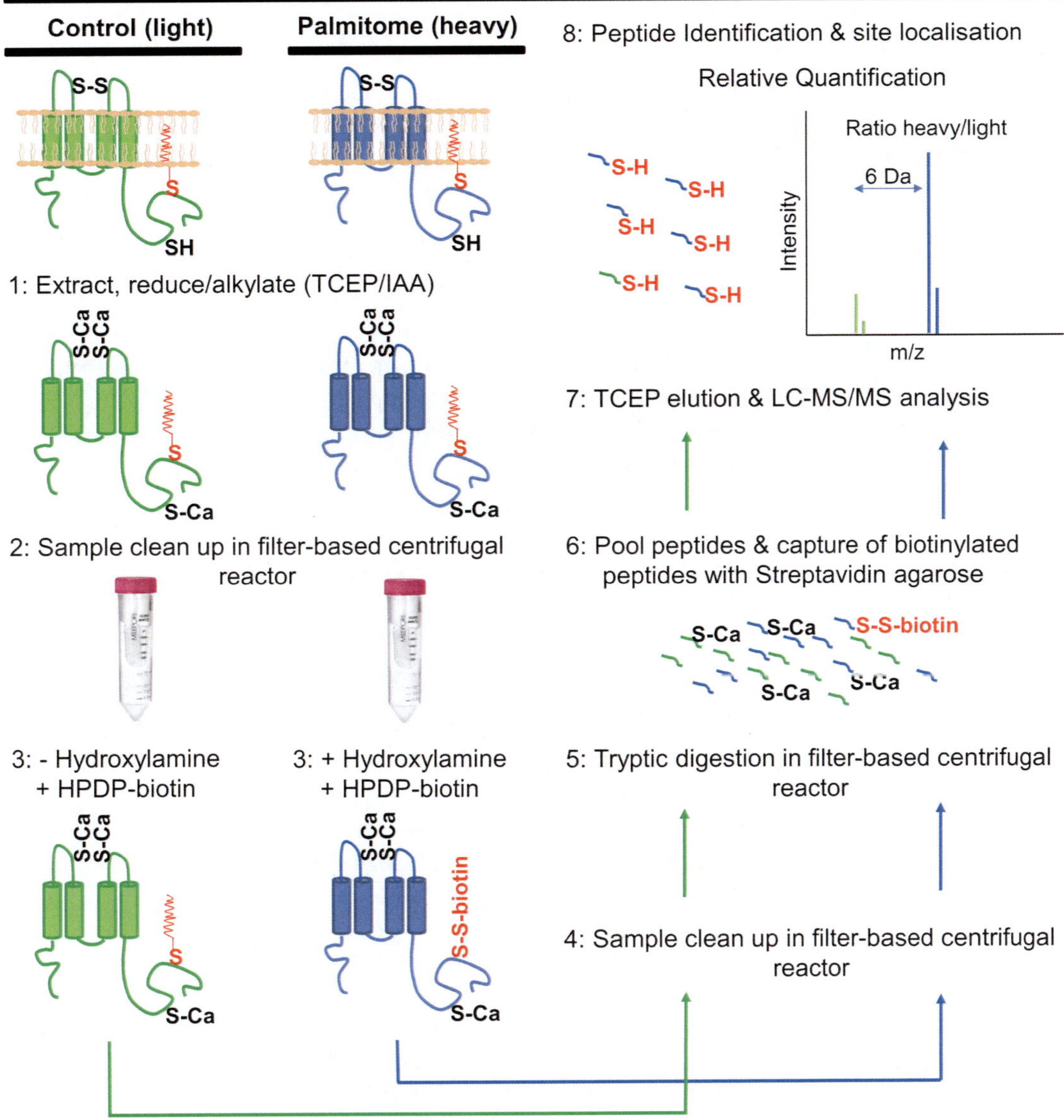

Fig. 1 Schematic representation of the ssABE workflow. Protein samples are (1) reduced and alkylated and transferred to a filter-based centrifugal unit in which the following steps are performed; (2) rapid removal of reagents, (3) release of palmitoyl groups from proteins by hydroxylamine treatment at neutral pH and concomitant biotinylation of these newly exposed cysteine residues with a TCEP-cleavable biotin (HPDP-biotin) (4) rapid removal of reagents, followed by (5) digestion of proteins into peptides and pooling of heavy and light peptides. (6) Biotinylated peptides are captured on Streptavidin agarose beads, washed and eluted (using TCEP) as free cysteine-containing peptides and (7) identified and quantified by LC-MS/MS analysis (8). Quantitative enrichment of peptides in Palmitome (plus hydroxylamine, (heavy)) versus Control (minus hydroxylamine, (light)) samples allows identification of previously palmitoylated peptides. Palmitoylation sites are localized in peptide sequences by the presence of non-alkylated (free) cysteine residues

2. Dounce homogenizer.
3. Heat block/thermomixer.
4. Fine gauge needle and syringe.
5. Tin foil to wrap samples in for alkylation (*see* **Note 2**).

2.2 Sample Cleanup

1. Amicon ultra 30 kDa molecular weight cutoff (MWCO) ultrafiltration centrifugal device or similar.
2. 8 M urea/100 mM Tris pH 8: add 100 mL of water to a volumetric flask and add 121.7 g of Trizma base. Add more water up to 800 mL, then adjust pH to 8 with hydrochloric acid and make up to 1 L with water to give a 1 M stock solution. Add 500 μL of this stock to a 50 mL tube, weigh out 24 g of urea and add to the tube. Make up to 50 mL with HPLC grade water (*see* **Note 3**). Store at room temperature. Prepare urea solution on the day of use.
3. Swinging-bucket centrifuge.

2.3 Acyl-Biotin Exchange

1. 50 mM Biotin-HPDP: dissolve 4.8 mg biotin-HPDP in 177 μL DMSO. Store 20 μL aliquots at −80 °C.
2. *N*,*N*-dimethylformamide .
3. 2 M hydroxylamine: dissolve 1.39 g of hydroxylamine hydrochloride in 5 mL of water. Titrate in 4 M sodium hydroxide until the solution reaches pH 7.4, then make up to 10 mL with water. Prepare immediately prior to use and keep on ice.
4. 8 M urea/100 mM ammonium bicarbonate: Add 5 mL of water to a 50 mL tube, add 24 g of urea and 0.395 g of ammonium bicarbonate and make up to 50 mL with HPLC grade water. Prepare on the day of use and store at room temperature (*see* **Note 4**).

2.4 Tryptic Digestion and Peptide Collection

1. 100 mM ammonium bicarbonate: Dissolve 0.395 g of ammonium bicarbonate in 50 mL of HPLC grade water. Prepare on the day of use and store at room temperature (*see* **Note 4**).
2. Trypsin: Resuspend 100 μg of lyophilized sequencing grade trypsin in 200 μL 0.1% trifluoroacetic acid. Store aliquots at −20 °C.
3. 2 × LB buffer: 100 mM Tris pH 7.4, 300 mM NaCl, 10 mM EDTA, 0.2% SDS and 0.4% Triton x-100. Store at room temperature.

2.5 Purification of Biotinylated Peptides

1. Streptavidin agarose resin.
2. 1 × LB buffer: 50 mM Tris pH 7.4, 150 mM NaCl, 5 mM EDTA, 0.1% SDS and 0.2% Triton x-100. Store at room temperature.

3. 2 M urea/Tris pH 7.4: add 5 mL of HPLC grade water to a 50 mL tube, add 6 g of urea, 500 μL of 1 M Tris pH 7.4 and make up to 50 mL with HPLC grade water (*see* **Note 3**).
4. 10 mM Tris(2-carboxyethyl)phosphine hydrochloride (TCEP).
5. 10% formic acid.
6. SpeedVac or another type of vacuum concentrator.

2.6 LC-MS/MS Analysis

1. Formic acid, LC-MS grade.
2. Ultrasonic bath.
3. High-resolution tandem mass spectrometer (e.g., Orbitrap) coupled to a nanoflow chromatography system fitted with a C18 trap column (100 μm, 75 μm × 2 cm, 5 μm) and a 50 cm C18 analytical column (2 μm, 75 μm id).

3 Methods

To be performed at room temperature unless stated.

The following protocol describes the application of ssABE with SILAC-based quantification to measure the enrichment of peptides in hydroxylamine-treated compared to control samples (Fig. 1). This quantitative enrichment allows the identification of putative sites of palmitoylation (Fig. 2). A variant of this methodology that permits relative quantification of palmitoylation site levels is described in Subheading 3.8 (Fig. 3).

3.1 Protein Extraction, Reduction, and Alkylation

1. Harvest SILAC-labeled (e.g., Lysine-8/Arginine-10) and unlabeled cells (5×10^6 cells is usually sufficient) in 1 mL extraction buffer. Homogenize samples 25 times with a Dounce homogenizer (*see* **Notes 5** and **6**). Transfer the sample to 2 mL tubes.
2. Heat at 50 °C for 20 min. Allow to cool slightly, then shear the DNA by passing the extract through fine gauge needle with a syringe ten times (*see* **Note 5**). Hold the needle aperture up against the side of the tube while syringing the sample to make sure the DNA is sheared effectively.
3. Remove any insoluble material by centrifugation at 20,000 × *g* in a benchtop centrifuge for 10 min and retain the supernatant.
4. Add additional iodoacetamide to bring the concentration to 50 mM and add urea to a final concentration of 8 M. Incubate the samples covered in tin foil or in the dark (*see* **Note 2**) for 3 h at room temperature with rotation to allow alkylation to proceed to completion.

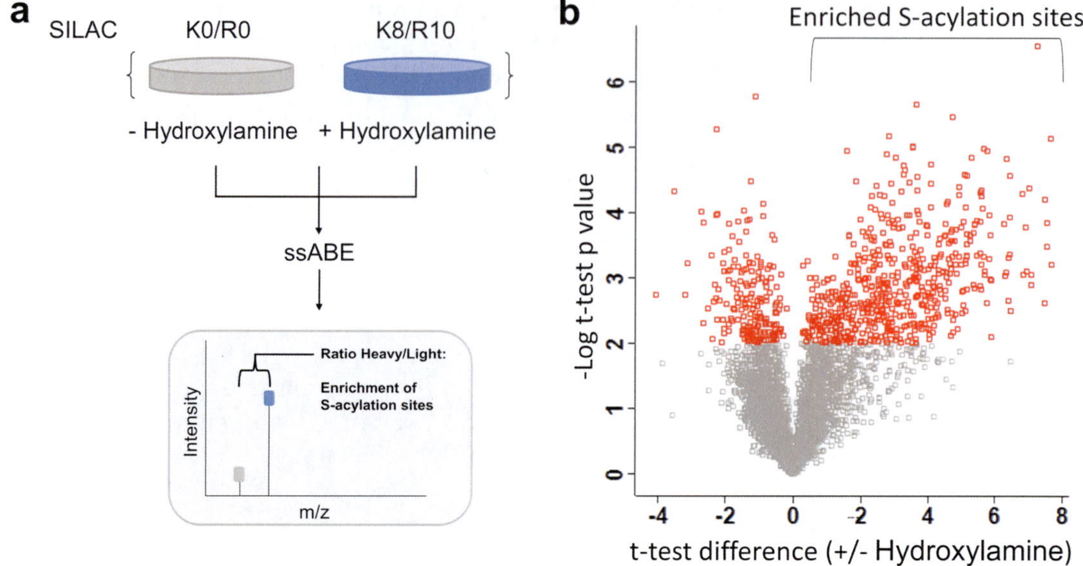

Fig. 2 Identification of putative *S*-acylation sites using ssABE and SILAC-based quantification. (**a**) A 2-plex SILAC experiment is performed where light (unlabeled) cells (−hydroxylamine) and heavy (Lysine 8 Arginine (K8/R10), labeled) cells (+hydroxylamine) are used for a ssABE workflow to identify *S*-acylation sites. (**b**) Volcano plot showing the quantitative enrichment of peptides in plus versus minus hydroxylamine-treated samples. Statistically significant hits (using t-testing and a permutation-based FDR) from this analysis are putative *S*-acylation sites

3.2 Sample Cleanup

1. Wash two 30 kDa MWCO spin columns with 8 M urea/100 mM Tris pH 8 and transfer an aliquot (2 mg each) of cell lysate (light and heavy) to each spin column and centrifuge for 10 min at 4000 × *g* to begin buffer exchange.
2. Buffer exchange the samples with the addition of 2 mL 8 M urea/100 mM Tris pH 7.4 and centrifugation for 10 min at 4000 × *g*, repeat five times (*see* **Note 7**). Transfer upper chambers to new 50 ml falcon tubes (*see* **Note 8**).

3.3 Acyl-Biotin Exchange.

1. Add 210 μL of 100 mM Tris pH 7.4 to the hydroxylamine-treated (+HA) sample (heavy SILAC labeled) and 560 μL to the control (−HA) sample (unlabeled).
2. Mix 4 μL of 50 mM HPDP-biotin with 20 μL *N*, *N*-dimethylformamide and add to both samples in the upper chamber of the unit to achieve a final concentration of 1 mM biotin.
3. Add 350 μL 2 M hydroxylamine (pH 7.4) to the +HA sample to a final concentration of 0.7 M. Allow hydroxylamine treatment and biotin labeling of newly released free cysteine residues to proceed for 1 h at room temperature with gentle agitation.

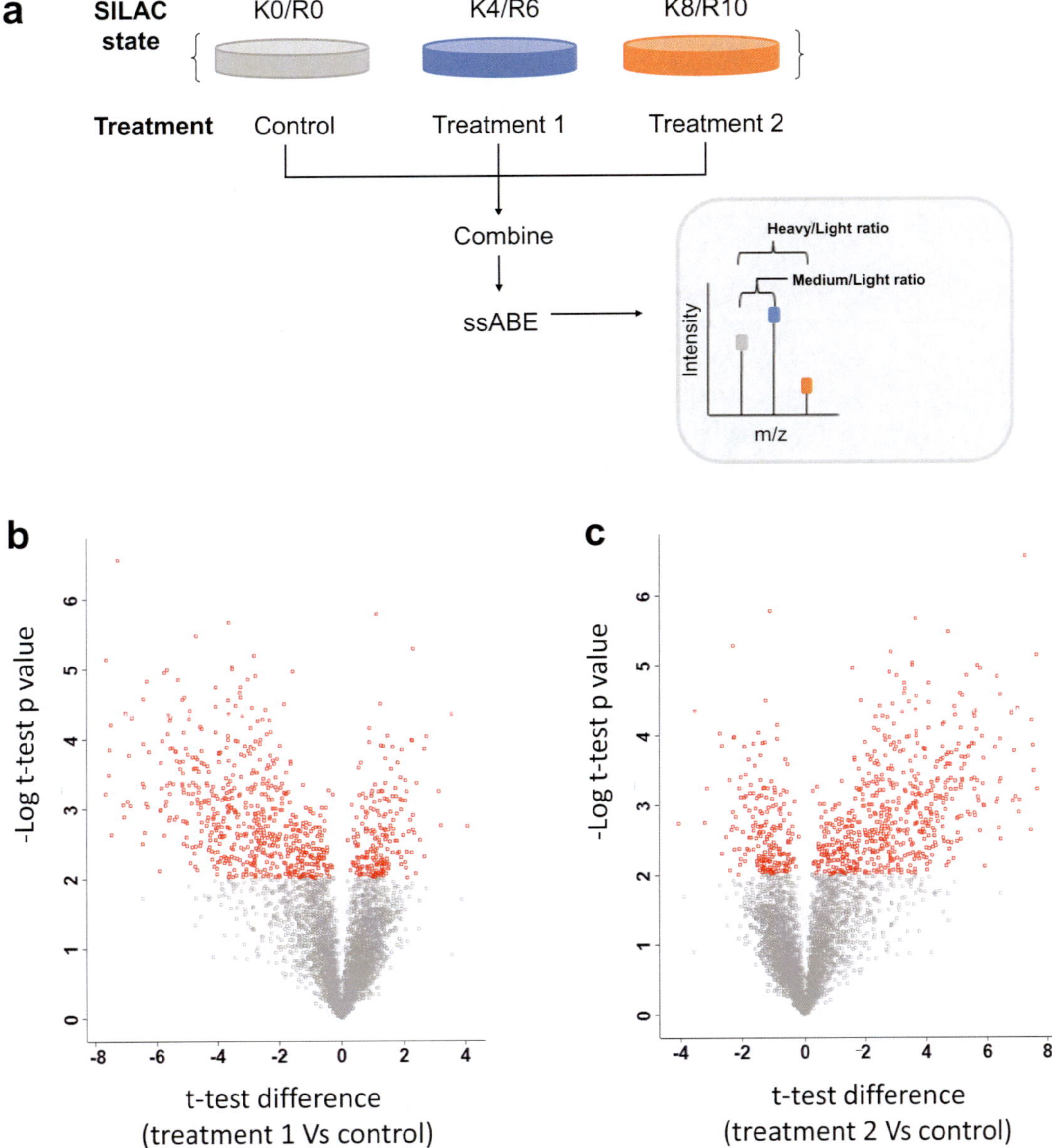

Fig. 3 Quantitative assay of global *S*-acylation site changes in perturbation experiments. (**a**) Schematic representation of a 3-plex SILAC experiment in which cells are either mock treated (light, K0/R0), or undergo two treatment/perturbation conditions (medium, K4/R8 and heavy K8/R10). Cell lysates are combined and are processed using the ssABE workflow with hydroxylamine treatment. (**b**, **c**) Volcano plots showing the quantitative differences between treatment 1 and control and treatment 2 and control. Differentially regulated *S*-acylation sites are identified using *t*-testing and a permutation-based FDR

4. Buffer exchange the samples with the addition of 2 mL 8 M urea/100 mM Tris pH 7.4 and centrifugation four times and a final wash with 2 mL 8 M urea/100 mM ammonium bicarbonate. Transfer upper chambers to new 50 ml falcon tubes (*see* **Note 8**).

3.4 Tryptic Digestion and Peptide Collection

1. Dilute the samples to a final concentration of 1 M urea by the addition of 100 mM ammonium bicarbonate. This is most easily achieved by adding 0.5 mL 100 mM ammonium bicarbonate, centrifuging to leave ~0.5 mL at 4 M urea then topping up to 2 mL with 100 mM ammonium bicarbonate.
2. Digest with sequencing grade trypsin at an enzyme-substrate ratio of 1:50 (40 μg trypsin in each sample) for 3 h at 37 °C with agitation.
3. Collect peptides from the upper chamber, place on ice, and wash the chamber membrane with 2 mL 2 × LB buffer.
4. After centrifugation, pool the collected peptides (final concentration of 0.5 M urea, 0.1% SDS, 0.2% Tx-100, 50 mM Tris pH 7.4, 150 mM NaCl, 50 mM ammonium bicarbonate) and store overnight at −20 °C (*see* **Note 9**).

3.5 Purification of Biotinylated Peptides

1. Wash 100 μL bed volume of Streptavidin-agarose resin three times with 1 mL of 1 × LB buffer, add to the pooled peptide samples (pooled light (control) and heavy (hydroxylamine treated)) and incubate for 1 h at room temperature.
2. Collect the beads by gentle centrifugation and wash the resin with 10 mL of 1 × LB buffer, incubating for 10 min with gentle agitation and repeat.
3. Transfer the resin to a mini-column and wash twice with 10 mL wash buffer (2 M urea/100 mM Tris pH 7.4) using a syringe, then with 2 mL HPLC grade water.
4. Elute peptides by the addition of 50 μL of 10 mM TCEP and incubate for 10 min at 37 °C. Repeat this step, pool the elutions, and acidify by the addition of 10% formic acid (*see* **Note 10**).
5. Dry the peptides down in a speed vac (*see* **Note 11**). Peptides can be stored at −20 °C.

3.6 LC-MS/MS Analysis

1. Resuspend dried peptides in 40 μL of 0.5% formic acid by pipetting up and down a number of times. Place samples in an ultrasonic bath for 5 min to ensure full resuspension.
2. Centrifuge samples at 20,000 × *g* for 5 min and transfer the supernatant (*see* **Note 12**) to an autosampler vial.
3. Samples are now ready for LC-MS/MS analysis. Ensure that the system is calibrated and quality control checks are performed, e.g., by analysis of a standard protein digest.
4. Peptides are desalted on a capillary trap column for 10 min at a flow rate of 5 μL/min and then separated using 120 min reverse phase gradient of 5–35% acetonitrile on the analytical column with a flow rate of 0.25 μL/min.

5. For a data-dependent analysis the mass spectrometer may be operated with a cycle of one MS acquired at high resolution, e.g. (60,000 at m/z 400), with the top 15 most abundant multiply-charged (2+ and higher) ions in a given chromatographic window subjected to fragmentation (e.g., collision induced dissociation, CID) (*see* **Note 13**).

3.7 Mass Spectrometry Data Analysis

1. Process raw MS data files using the data analysis platform MaxQuant [9] or equivalent, searching the data against an appropriate up-to-date species-specific UniProt sequence database.
2. Set up the search using the following parameters: trypsin as the digesting enzyme with a maximum of two missed cleavages, 7 ppm for MS mass tolerance, 0.5 Da for MS/MS mass tolerance (these settings are mass spectrometer dependent), with acetylation (Protein N-term) and oxidation (M) as variable modifications and carbamidomethyl (C) as a fixed modification.
3. Palmitoylation sites are reported and localized in peptide sequences by the use of a "loss" of 57.02146 Da (i.e., no carbamidomethyl modification/unmodified cysteine) as a variable modification (*see* **Note 14**) as only those cysteines targeted by acyl-biotin exchange should be unmodified, all other cysteines should be alkylated.
4. A protein False Discovery Rate (FDR) of 0.01 and a peptide FDR of 0.01 should be used for identification level cutoffs.
5. Load the "Palmitoylation sites" txt file generated by MaxQuant into Perseus.
6. Remove "reverse" and "contaminant" identifications.
7. In order to identify confidently assigned sites of palmitoylation in peptide sequences, remove those identifications with a localization score of <0.75 and a score diff <5.
8. Filter the data so that only putative palmitoylation site identifications with at least 3 valid values (i.e., 3 measured peptide ratios) across biological replicates in the hydroxylamine-treated samples and log2 transform these values.
9. Check that the correlation of peptide ratios between replicates is high and that the data is normally distributed.
10. Peptide ratios calculated by MaxQuant are then used to quantify enrichment of peptides in +/− hydroxylamine-treated samples across replicates. A minimum of three biological replicates is required.
11. Perform a one-sample test with a Benjamini-Hochberg FDR calculation in Perseus. Peptides within an FDR of 0.05 and an

S0 value = 1 are statistically enriched in the hydroxylamine treated versus control.

12. In order to increase the stringency of analysis and additional fold change enrichment filter should be applied (at least a threefold change).
13. The resultant set of sites is defined as being quantitatively enriched in the hydroxylamine-treated samples and therefore peptides with putative sites of palmitoylation. *See* **Notes 15** and **16** for identification of false positives.
14. Further validation of sites can then be performed using techniques such as site-directed mutagenesis and PEG switch methods [10].

3.8 Alternative Workflow for Relative Quantification of S-Acylation Sites in Perturbation Experiments

Once the repertoire of palmitoylation sites expressed in the system of interest has been characterized using the standard ssABE protocol described above, a variation of this procedure may be used to monitor changes in palmitoylation levels in perturbation experiments.

1. In order to accurately compare palmitoylation levels between control (untreated) and treated conditions, 2-plex (1 treatment condition) or 3-plex (2 treatment conditions) cell lysates from SILAC-labeled cells are pooled at the beginning of the modified ssABE protocol (Fig. 3).
2. The pooled sample is reduced/alkylated and treated with hydroxylamine and processed and analyzed as a single sample. This allows for precise quantitation of palmitoylation levels between conditions.
3. If the treatment duration is likely to cause changes in protein expression (e.g., gene knockout, knockdown, or chronic drug treatment), then global protein expression profiling of the systems will also need to be performed.
4. Palmitoylation site levels may then be normalized to protein expression levels to reflect true changes in palmitoylation. However, if the treatment duration is acute, then expression profiling is not required.

4 Notes

1. Instead of PMSF, leupeptin, and aprotinin, a protease inhibitor cocktail can be used at appropriate dilutions.
2. Samples need to be wrapped in tin foil to protect them from light as iodoacetamide is photosensitive.

3. Solvation of urea is an endothermic process, so it can take some time. Placing the tube in a warm water bath (<37 °C) will speed up the process.
4. Ammonium bicarbonate needs to be prepared fresh as it breaks down leading to a loss of buffering capacity of the solution.
5. To avoid bubbles/foam in the samples, the plunger of the homogenizer should stay below the level of the liquid. Similarly, excessive foaming can occur during sheering of DNA if the entire sample is drawn up into the syringe, so it is best to leave some of the sample in the tube when sheering DNA.
6. The sample will be very viscous after homogenization due to the DNA content, so it can be difficult to pipette during transfer to the tubes.
7. Volume after each centrifugation step should be around 0.5 mL in the upper chamber. It can take more than one centrifugation step to fully remove all the liquid in the upper chamber to achieve a 0.5 mL volume.
8. The collection tube which the filter unit fits into will need to be changed multiple times to avoid the waste liquid that has passed through it contaminating the filter unit during the biotinylation and digestion steps. The unit fits well into a standard 50 mL falcon tube.
9. An aliquot (20 μL) of the sample can be taken at this point for LC-MS/MS analysis to ensure the digestion step has been successful.
10. Ensure all solutions used in this step are made up with HPLC grade water to prevent any potential interference in the LC system by contaminants.
11. Do not overdry the peptides as it can be difficult to resuspend them prior to analysis.
12. It is important to avoid the transfer of any of the pellet at this point as any insoluble material could block either the autosampler needle or the HPLC trap column.
13. Acquisition parameters should be adjusted according to the type of instrument used. MS scans should be acquired with high resolution for SILAC-based quantification.
14. An "unmodified cysteine" modification is not available as standard in the MaxQuant package but can be added in the Andromeda configuration by specifying a modification with the following composition: H(−3) O(−1) C(−2) N(−1) (Δ −57.02146 Daltons) and a cysteine specificity.
15. A number of known false positives can be found as positive hits for palmitoylation during the downstream analysis. A major source of false positives can arise from incomplete alkylation of

cysteine residues; the extent of alkylation should be checked. In addition, enzymes which bind biotin and enzymes which use thioester intermediates in their reaction mechanism are known false positives. Such proteins will need to be removed from the dataset before further data analysis.

16. It is important to cross-reference any putative sites of palmitoylation with known disulfide bonds. While it is not impossible that a cysteine residue could be palmitoylated and involved in a disulfide bond depending on the context, there is no way of knowing if this is the case or the positive hit is due to incomplete disulfide reduction and blocking of the resulting cysteine.

References

1. Magee AI, Gutierrez L, McKay IA, Marshall CJ, Hall A (1987) Dynamic fatty acylation of p21N-ras. EMBO J 6(11):3353–3357
2. Topinka JR, Bredt DS (1998) N-terminal palmitoylation of PSD-95 regulates association with cell membranes and interaction with K+ channel Kv1.4. Neuron 20(1): 125–134
3. Blanc M, David F, Abrami L, Migliozzi D, Armand F, Burgi J et al (2015) SwissPalm: protein palmitoylation database. F1000Res 4:261
4. Drisdel RC, Green WN (2004) Labeling and quantifying sites of protein palmitoylation. BioTechniques 36(2):276–285
5. Forrester MT, Hess DT, Thompson JW, Hultman R, Moseley MA, Stamler JS et al (2011) Site-specific analysis of protein S-acylation by resin-assisted capture. J Lipid Res 52(2):393–398
6. Martin BR, Cravatt BF (2009) Large-scale profiling of protein palmitoylation in mammalian cells. Nat Methods 6(2):135–138
7. Jones ML, Collins MO, Goulding D, Choudhary JS, Rayner JC (2012) Analysis of protein palmitoylation reveals a pervasive role in Plasmodium development and pathogenesis. Cell Host Microbe 12(2):246–258
8. Collins MO, Woodley KT, Choudhary JS (2017) Global, site-specific analysis of neuronal protein S-acylation. Sci Rep 7(1):4683
9. Cox J, Mann M (2008) MaxQuant enables high peptide identification rates, individualized p.p.b.-range mass accuracies and proteome-wide protein quantification. Nat Biotechnol 26(12):1367–1372
10. Percher A, Ramakrishnan S, Thinon E, Yuan X, Yount JS, Hang HC (2016) Mass-tag labeling reveals site-specific and endogenous levels of protein S-fatty acylation. Proc Natl Acad Sci U S A 113(16):4302–4307

Chapter 7

Reducing Complexity? Cysteine Reduction and *S*-Alkylation in Proteomic Workflows: Practical Considerations

Caroline A. Evans

Abstract

Reduction and alkylation are common processing steps in sample preparation for qualitative and quantitative proteomic analyses. In principle, these steps mitigate the limitations resulting from the presence of disulfide bridges. There has been recurring debate in the proteomics community around their use, with concern over negative impacts that result from overalkylation (off-target, non-thiol sites) or incomplete reduction and/or *S*-alkylation of cysteine. This chapter integrates findings from a number of studies on different reduction and alkylation strategies, to guide users in experimental design for their optimal use in proteomic workflows.

Key words N-terminal alkylation, Alkylation, Reduction, Cysteine, Artifact, Proteomic

1 Introduction

Proteomics provides a core platform for the identification, characterization, and quantification of proteins either at the level of intact proteins or proteolytic fragments via "top-down" or "bottom-up" mass spectrometry (MS) based approaches. Sample preparation protocols, downstream of protein extraction and solubilization, typically include reduction of disulfide bonds followed by the *S*-alkylation or "capping" of reduced cysteine thiols. Cysteine is one of the most susceptible residues for posttranslational modifications (PTMs), undergoing a range PTMs, which include disulfide bonds linking cysteine residues to form intra- and inter-molecular bridges present in the same or different polypeptide chains. Disulfide bonds contribute to tertiary structure associated with protein folding [1] being dynamically interchangeable dependent on the redox state [2]. The addition of reducing agent disrupts disulfide bridges to unfold proteins and dissociate protein samples into the component polypeptides. Alkylating agents prevent the regeneration of native

Caroline A. Evans et al. (eds.), *Mass Spectrometry of Proteins: Methods and Protocols*, Methods in Molecular Biology, vol. 1977, https://doi.org/10.1007/978-1-4939-9232-4_7,

disulfide bridge(s) and, critically, also prevent the formation of non-native, random "scrambled" linkages to mitigate complications arising from this type of disulfide linkages [3, 4].

In top-down approaches, reduction and alkylation of cysteines are required to ensure that individual proteoforms can be resolved to single species [5]. Bottom-up approaches additionally benefit from (i) the increased efficiency of proteolytic enzymes afforded by improved access to cleavage sites and (ii) the detection of cysteine-containing peptides [6, 7] (*see* **Notes 1** and **2**). *S*-alkylation strategies also find application in the analysis of cysteine PTMs (both redox and non-redox), using the biotin switch assay approach in which the labile PTM sites are replaced with a stable biotin moiety to facilitate their enrichment, reviewed by Wojdyla and Rogowska-Wrzesinsk [8]. A variant of this technique, site specific Acyl-Biotin Exchange, is detailed in the accompanying chapter by Woodley and Collins as a method for the identification and the quantitative profiling of *S*-palmitoylation employing SILAC.

For bottom-up approaches, the proteolytic digestion is performed either using gel-separated proteins or in-solution digests of simple or complex samples: from single proteins to whole cell or tissue proteomes. In practical terms, the simplest gel-based method is 1D SDS-PAGE. Proteins are resolved based on molecular weight and the method offers the advantage of compatibility with many of the detergents and chaotropes required for protein extraction protocols. The one-step sample cleanup and protein separation afforded by SDS-PAGE has resulted in its wide application to analysis of simple and complex mixtures, either from individual gel bands or, in the case of GeLC-MS/MS, entire gel lanes for whole sample analysis [9]. Higher protein resolution results from the use of isoelectric focusing prior to PAGE. Proteins are thus resolved based on two properties, charge and molecular weight by two-dimensional gel electrophoresis (2D-PAGE). The gel spots and gel bands from 1D and 2D PAGE gels are excised and proteolytically digested "in gel" in preparation for MS analysis [10], typically using trypsin alone or in combination of LysC. Proteins can also be digested in solution rather than in gel. Alternative enzymes such as LysN, AspN, GluC, and ArgC can also be used, particularly for the analysis of PTMs, which can result in missed cleavage by trypsin [11].

Reduction and alkylation workflows have been subject to debate from the dawn of proteomics. There are two main questions related to this in terms of protocol: When to reduce and *S*-alkylate? How to limit artifactual chemical modifications from these processing steps? (*see* **Note 3**). These questions are directed to the requirement that sample content reflects biological complexity rather than experimental processing artifacts. This concern is in common with all aspects of sample preparation for proteomic analysis where the introduction of artifacts and contaminants is avoided in the design and practice of proteomic workflow as reviewed by Bittremieux et al. [12]. Studies systematically comparing reduction and alkylation protocols indicate that balancing maximum digestion efficiency with minimum off-target modifications resulting from side reactions of reduction and alkylation poses a significant challenge.

The aim of this chapter is to signpost to key publications relating to reduction and alkylation protocols, to integrate the main findings, and to provide a user guide to experimental design considerations for inclusion of optimal reduction and *S*-alkylation protocols. Tables 1 and 2 list commonly used reducing and alkylation reagents with notes on their application. Key publications relating to critical evaluation of reduction and alkylation workflows are discussed, particularly those directed to comparing alternative to iodoacetamide, the most widely used alkylation reagent. Key findings are summarized in Table 3. Design considerations for application to proteomic workflows are provided in Notes, Subheading 3.

2 Reduction and Alkylation Workflows in Gel Electrophoresis—Historical Perspective and Update

Reduction and *S*-alkylation have been included in protein analysis from early experiments using 2D gels through to later development of gel free methods. The requirement for and the use of reduction and *S*-alkylation protocols in 2D PAGE analysis were evaluated and critically reviewed in a series of papers in the early 2000s [4, 13, 14]. Denaturing IEF requires cysteine reduction in order to separate and unfold polypeptide chains, using immobilized pH gradient (IPG) strips, with thiol-reduced proteins loaded onto the strip. During IEF, the negatively charged reducing agent migrates from and thus depletes in the basic region leading to (re) oxidation and S–S bridge formation. Inclusion of an alkylation step, prior to IEF, prevents resulting artifacts: (i) blurred or comet-shaped spot streaks, spot appearance/disappearance that result from homo- and hetero-oligomers, generated by the formation of scrambled disulfide bridges [4] and (ii) degradation due to beta-elimination of the –SH groups of cysteine leading to conversion to dehydroalanine and hydrolysis of adjacent peptide bonds [4, 13]. In top-down analysis, success requires reduction and *S*-alkylation of cysteine to ensure each proteoform resolves to the correct pI and molecular weight on 2D PAGE. For more information, Padula et al. provide a comprehensive guide to sample preparation and top-down protein analysis [5].

Despite the benefits of alkylation post reduction, typical 2D workflows only reduce proteins prior to IEF. Reduction and alkylation reagents, usually dithiothreitol and iodoacetamide, are included in the two-step equilibration process for exchange of denaturants prior to second-dimension SDS-PAGE to remove the disulfide

Table 1
Commonly used reducing agents and notes on their use

Reducing agent	Notes
Dithiothreitol (DTT)	Most commonly used thiol reducing agent Optimal pH range 7.1 and 8.0 More stable and less toxic alternative to beta-mercaptoethanol, pH Can be used to quench alkylation [26], *see* **Note 4** To maximize performance and minimize unwanted off target alkylation, usage of DTT is recommended in combination with acrylamide (AA) for in-solution digests [7] Can form cysteine adduct of mass addition +151.996571 Da Ubiquitin may form two spots in 2D gels when a DTT-containing buffer is used [45]
Tris-2(-carboxyethyl)-(TCEP)	Less susceptible to oxidation than DTT and odorless (unlike DTT, BME), works at low pH and wider pH range Superceded tributylphosphine (effective but toxic and low solubility in water solutions) [46] Commonly used in conjunction with methylmethanethiosulfonate (MMTS) for iTRAQ workflows Reduction and alkylation can be achieved in a single step using TCEP as the reductant since phosphines do not quench alkylation reagents [7, 46] TCEP less stable in phosphate buffers, especially at neutral or alkaline pH, therefore prepare and use freshly made TCEP stock solution Applicable to Immobilized Metal Affinity Capture workflows for PTM analysis since TCEP does not reduce metal ions Residual TCEP can interfere with Isoelectric Electrophoresis Focusing and side reactions reported including (i) Reduction of oxidized methionine residues present on the protein prior to digestion. Avoid for experiments where monitoring methionine oxidation, such as monitoring of protein therapeutic [3, 34] (ii) Cleavage of cysteine residues to form heterogeneous termini are formed, similar to those occurring due to cysteine cyanylation [47] (iii) Conversion of cysteine into dehydroalanine via β-elimination of a disulfide bond [48] Immobilization of TCEP on magnetic carbon-coated Cobalt nanoparticles, provides a simple and rapid approach to its removal by use of a magnet [49]
Tris(hydroxypropyl) phosphine (THP, THPP)	Volatile. Stable and resistant to ambient oxidation Can be used in conjunction with iodoethanol, excess reagent can be removed under vacuum For gel slices, the procedure combines washing, destaining, reduction and alkylation into a single step [50] Utilized for label free phosphoproteomic analysis in the accompanying chapter by Barrios-Llerena and Le Bihan
Beta mercaptoethanol (2-ME or βME, BME, ME)	A thiol compound, more stable than DTT, typically used as the reducing agent in the Laemmli sample buffer for 1D SDS polyacrylamide gel electrophoresis Using BME with acrylamide has been recommended for in-gel digests [7] Potential to form an artifact ME (Cys)S-S-C_2H_4OH 7 artifact which results in mass addition of 75.9983. Use is not recommended for analysis of protein modifications [44]

Table 2
Commonly used alkylating agents, their abbreviations in the literature and notes on their use

Alkylating reagent	Adduct (as listed by UNIMOD, unimod.org [35]	Monoisotopic mass addition (Da)	Notes
Iodoacetamide (IAM, IAA)	Carbamidomethyl H(3) C(2) N O	57.021464	Light sensitive, halo(iodo)alkane Significant off target can occur in addition to *S*-alkylation (*see* Subheading 2.1) For analysis of ubiquitin: chloroacetamide is recommended instead. The 2-acetamidoacetamide covalent adduct to lysine is isobaric with diglycine (114.0429) [38]
Iodoacetic acid (IAA, IAC)[a]	Carboxymethyl H(2) C(2) O(2)	58.005479	Off target similar to the related iodoalkane, iodoacetamide been noted [26, 27] Introduces a negative charge
Chloroacetamide (CA, CAA)	Carbamidomethyl H(3) C(2) N O	57.021464	Light sensitive, halo(chloro)alkane Same mass addition as iodoacetamide CA has the same efficacy as IAM as an alkylating agent but results in virtually no off-target alkylation. Side reactions of methionine and tryptophan oxidation [24] Suggested alternative to iodoacetamide for analysis of ubiquitin linkage [38]. It should be noted, that neutral loss of one or both glycines of the ubiquitin diglycine footprint enables it to be distinguished from dicarbamidomethylation [51]
Acrylamide (AA)	Propionamide H(5) C(3) N O	71.037114	Proposed as a replacement to iodoacetamide for in-solution and in-gel digests [7] Can be used for in situ alkylation of proteins during 1D and 2D PAGE [17] Gel-assisted sample preparation (GASP) co-polymerizes and alkylates samples with acrylamide [20] Acrylamide together with deuterated (D3)-acrylamide has been used for relative quantification [19]
Methyl methanethiosulfonate (MMTS)	Beta-methylthiolation H(2) C S	45.987721	Used in conjunction with TCEP for iTRAQ and related workflows
N-ethylmaleimide (NEM)	*N*-ethylmaleimide on cysteines H(7) C(6) N O(2)	125.047679	N terminal alkylation, lysine alkylation, alkylation of histidine in addition to *S*-Alkylation [36] No evidence for hydroxyl or thioether alkylation was observed [25, 36]
4-Vinylpyridine	*S*-pyridylethylation H(7) C(7) N	105.057849	An alternative alkylating reagent, determined to be less effective than acrylamide for in-solution digests [25]

[a]Note that IAA and IAM have been used as an abbreviation for iodoacetamide and iodoacetic acid, dependent on the publication and thus iodoacetamide and iodoacetic acid are not abbreviated in the main text

Table 3
Key publications comparing different alkylating reagents relative to iodoacetamide

Authors	Study	Key results	Conclusions
Müller and Winter [7]	Compared common reduction reagents (DTT, TCEP, and BME) and alkylation reagents (iodoacetamide and related iodoalkane, iodoacetic acid; acrylamide, and chloroacetamide, selected as non-iodine alternatives) *In-solution digest conditions:* Buffer—0.1 M Tris–HCl/4% SDS Reduction—20 mM DTT, 20 mM TCEP, or 40 mM BME for 30 min at 56 °C Alkylation—20 mM each of the individual alkylation reagents, 30 min at 23 °C in the darkSample quenching by addition of DTT, TCEP, or BME *In-gel digestion conditions:* Buffer—0.1 M NH_4HCO_3 Reduction—20 mM DTT, 20 mM TCEP, or 40 mM BME for 45 min at 56 °C Alkylation—55 mM each of the individual alkylation reagents, 30 min at 23 °C in the dark	Greatest variability resulted from alkylation rather than reduction. Alkylation reagents tested have similar cysteine-alkylation efficiencies but iodoacetamide and iodoacetic acid have more off-target effects. Acrylamide identified as the best *S*-alkylation agent Differences between in-solution and in-gel digestion—less off-target alkylation and higher efficiency of *S*-alkylation for in-solution digests relative to in-gel digests Alkylation of methionine by iodoacetamide, iodoacetic acid (iodoalkane reagents) causes significant reduction in the number of methionine-containing peptides	Avoid iodoalkanes, replace with acrylamide Acrylamide best used in combination with the DTT reducing agent for in-solution digestion protocols Acrylamide best used in combination with BME reducing agent for in-gel digestion protocols Reduction and alkylation of proteins *pre* tryptic digestion with FASP reduces off-target artifacts—*see* **Note 5**. Comprehensive analysis of the benfits and drawbacks resulting from reduction and alkylation protocols.
Hain and Robinson [24]	Comparison of iodoacetamide and chloroacetamide as alkylating agents prior to in trypsin or LysC-trypsin digestion and subsequent LC–MS/MS analysis *In-solution digest conditions:* Buffer—6 M Urea in 50 mM NH_4HCO_3, pH 7.8, sample prepared with a Barocycler at 35 °C. Reduction—10 mM TCEP. Alkylation—40 mM iodoacetamide or chloroacetamide (referenced protocol of Rebecchi et al. [28])	Similar efficiency for *S*-alkylation Chloroacetamide has lower off-target alkylation relative to iodoacetamide, particularly noted for N-terminal alkylation occurrence of 0.05%, tenfold lower than for iodoacetamide Higher levels of oxidation of Met (40%) and Tryptophan (mono and di-oxidation) relative to iodoacetamide	Lower off-target alkylation than iodoacetamide however an adverse impact of 2-chloroacetamide on methionine oxidation reported. Confirmed by use of standard peptides to evaluate reduction and alkylation efficiency and specificity [39]

Suttapitugsakul et al. [25]	Systematically compared: Reduction of the disulfide bonds with commonly used DTT, 2-ME, TCEP, or tris(3-hydroxypropyl)phosphine (THPP). 5 mM, incubation 56 °C for 25 min Subsequent alkylation with iodoacetamide, acrylamide, *N*-ethylmaleimide, or 4-vinylpyridine. 14 mM, incubated at room temperature for 30 min, in the dark. Quenched by addition of the same amount of each reducing reagent	Greatest variability resulted from alkylation rather than reduction Iodoacetamide was optimal for cysteine alkylation, considering reaction completion rate and off target. The use of acrylamide resulted in similar data	Iodoacetamide and acrylamide recommended for *S*-alkylation. Acrylamide provides an alternative to iodoacetamide
Huang et al. [37]	Positional proteomics approach: enzyme and chemical assisted N-terminal blocked peptides analysis, termed ENCHANT Buffer—100 mM TEAB and 4 M guanidine hydrochloride. Reduction—10 mM DTT at 37 °C for 45 min. Alkylation—30 mM iodoacetamide in dark (Shotgun and ENCHANT approaches), or 100 mM acrylamide (Nα-acetylome approach) for 1 h at room temperature	Side reaction ratios in Nα-amine and other possible amino residues were >0.5% for both iodoacetamide and acrylamide	Limited off target effects for acrylamide and iodoacetamide in this study. Proposed to be due to use of FASP for the in-gel digestion approach, which would efficiently remove excess alkylating reagents

Concentrations listed for the reducing and alkylating reagents are the final concentrations bonds formed during IEF. This is however less effective than performing reduction and alkylation before IEF [4, 15]. The issue was revisited in 2017, in an article entitled "Reduction and Alkylation of Proteins in 2D Gel Electrophoresis: Before or after Isoelectric Focusing?" concluding that it is debatable whether the reduction and alkylation of proteins between the two dimensions of 2DE is required, when already performed at the pre-IEF stage [16]. As noted in 2001, in principle, addition of reducing agent post-IEF is too late, particularly since acrylamide monomers (predominantly from SDS-PAGE, but potentially from IPG strips) can form adducts on free cysteine thiols [4]. The use of acrylamide as the *S*-alkylation reagent is recommended since it both prevents protein degradation and avoids "mixed" alkylation products arising from use of more than alkylation reagent [4, 13]. Acrylamide is an effective *S*-alkylation reagent in gel electrophoresis protocols [17].

It has also been recommended to reduce and alkylate protein samples before SDS-PAGE [18]. In practice, while proteins are often prepared under reducing conditions, many users do not combine reduction and alkylation before SDS-PAGE, instead performing these at the stage of in-gel digestion [7]. It has been estimated that there is 30 mM of free acrylamide in the gel matrix, which risks *S*-alkylation by acrylamide to form propionamide adduct(s), when samples are reduced but not alkylated prior to electrophoresis. Addition of 1% acrylamide prior to loading proteins onto the gel completes the reduction and alkylation steps at this stage [19]. A "gel-assisted" method for proteomic sample preparation utilizes alkylation by acrylamide to remove the need for alkylation as an additional processing step. Solubilized protein extracts or intact cells are co-polymerized with acrylamide, facilitating denaturation, reduction and cysteine alkylation in a gel matrix. Samples in this format can be directly subject to proteolytic digestion [20].

Debate continues on reduction and alkylation, in part due to renewed interest in 2D gels, with respect to analysis of proteoforms for top-down workflows. 2D PAGE is well suited for resolution of species of differing isoelectric point and molecular weight, a critical requirement for resolution of individual protein sequence isoform and posttranslationally modified protein variant(s) to single species [21, 22]. For bottom-up proteomics, debate has been renewed as a result of systematic evaluation of the commonly used protocols for reduction and alkylation protocols in LC-MS/MS analyses [7, 23–25], raising questions as to whether current protocols are best suited and as to what changes could improve current practice for sample processing in gel or in solution.

2.1 Reduction and Alkylation Reagents

Controlling the chemical reactions associated with reduction and *S*-alkylation is critical for qualitative and quantitative analyses: incomplete or off-target modifications will impact data quality and introduce inter-sample variability. Since it is difficult to estimate the number of cysteines in (often complex) samples, proteins are usually processed with an excess of reducing and alkylating reagents [26–28]. The thiol compound, dithiothreitol (DTT) and phosphine, tris(2-carboxyethyl)phosphine (TCEP) are most commonly used for reduction [7]. TCEP is odorless and more stable than DTT. TCEP is irreversible reducing agent that, unlike DTT, cannot directly catalyze thiol-disulfide exchange reactions. Alkylation chemistry is typically achieved by two mechanisms of bimolecular nucleophilic substitution (by haloalkanes) or Michael addition (non haloalkanes) using alkylating agents as Michael acceptors. Tables 1 and 2 list reduction and alkylating reagents, respectively, with notes on their use in proteomic workflows.

Direct comparison of reducing and alkylation reagents indicates that the greatest variability arises from choice of reagent for alkylation [7, 25]. Under-alkylation and overalkylation of non-thiol residues are two key issues that were identified early in the development of proteomic analysis [26]. Under-alkylation can occur if free cysteine thiols are not available, potentially due to incomplete protein folding or incomplete reduction of disulfide bonds. *S*-alkylation may be incomplete if the reaction time is not sufficient; and if too long, then off-target (also termed off site) reactions can occur.

Iodoacetamide was first reported to result in N-terminal alkylation [26]; this modification is also reported for acrylamide, *N*-ethylmaleimide, or 4-vinylpyridine [25]. It has become clear that additional alkylation events occur from bimolecular nucleophilic substitution reactions by the electron-rich groups of the side chains of non-cysteine amino acids [25]. Off-target or side reactions occur using iodoacetamide with alkylation reported for "cysteine-less" peptides [27] including *N*- and *O*-alkylation of N-terminal amine and C-terminal carboxylic acid groups, respectively [29]. Additional alkylation events have been reported: *O*-alkylation of hydroxyls in tyrosine, aspartate, glutamate, serine and threonine [7, 23], *N*-alkylation of free *epsilon* amino group of lysine residues with lower propensity than alkylation of free N-terminal amines [23], alkylation of the thioethyl group of methionine [30, 31] and alkylation of the imidazole nitrogens of histidine [23, 32]. Alkylation can occur singly or cooperatively among different amino acid residues in the same peptide [23]. *N*-alkylation of lysine side chains results in mis-cleavage by trypsin due to steric hindrance [33]. Off-target can be limited by quenching with thiols [26] (*see* **Note 4**). Incomplete or off-target modifications impact data quality in qualitative and quantitative workflows. This is due to increased sample complexity resulting from artifacts so that multiple, differently modified forms of each peptide can exist. This is an issue for both characterization of a single protein, for example, a protein biosimilar [3, 34], and profiling of complex samples [7]. In addition, accurate quantification requires that the quantotypic peptide (stoichiometric to the amount of corresponding protein) is present as a single species.

While the presence of peptide variants can be detected using data-independent approaches to fragment all peptides, failure to include these peptides m/z values in targeted precursor or precursor-product transition mass lists results in their omission from the analysis. It may be possible to account for the additional reduction/alkylation related variants, allowing for all possible forms in downstream database searching and matching of peptide spectral masses, but this is not without consequence in terms of increased database search space, processing time and false discovery rate (FDR) scoring. The presence of the modified forms also increases sample complexity for HPLC methods and MS analysis, leading to reduced sensitivity [12].

2.2 Evaluation of Current and Alternative Reduction and Alkylation Protocols

In terms of assessing data quality and alkylation specificity, it has been recommended to include *N*-alkylation at the N terminus in database searching, in addition to *S*-alkylation [27]. Comparison of commonly used reduction and alkylation protocols has provided key findings to inform user choice of protocols and detail the prevalence of off-target products. Table 3 provides details on key publications in this area, which relate to critical evaluation of current practice and alternatives for the analysis of complex samples, as analyzed in large-scale studies. A systematic study compared three reducing agents (DTT, TCEP, and BME), each tested in combination with iodoalkanes (iodoacetamide, iodoacetic acid) and non-iodine-containing alkylation agents, chloroacetamide (a chloroalkane) and acrylamide. These reflect the most commonly used reduction and alkylation reagents, based on a survey of the literature [7]. The experiment tested the protocols for in-solution digests and in-gel digests, using complex samples, reflecting current practice in high-throughput bottom-up proteomic approaches. A major overall conclusion of the study by Müller and Winter confirmed prior findings that iodoacetamide has the potential to modify all reactive groups of amino acids and signals caution in its use [7]. Iodine-containing alkylation reagents resulted in off-target at peptide N-terminus followed by lysine, glutamic acid, and histidine residues in terms of occurrence. Acrylamide and chloroacetamide resulted in similar numbers of off-target alkylation, with the N-terminus, serine, threonine, and glutamic acid modified most frequently [7]. The benefits and drawbacks of each method were evaluated.

A striking difference between alkylation reagents was that the lowest efficiencies of peptide identification result from the use of iodoacetamide and iodoacetic acid for both in-gel and in-solution digestion [7]. These iodine-containing alkylation reagents led to an apparent loss of methionine-containing peptides. An elegant set of experiments using dimethyl labeling, mass-tolerant searches, and synthetic peptide provided clear evidence that the missing methionine peptides occurred as result of side chain loss from alkylated methionine (-SHCH3, -48 da) occurring during MS/MS analysis. The side chain loss may occur during electrospray ionization or in MS/MS fragmentation [7], listed as "Dethiomethyl" in the Unimod database [35].

Alternative, non-iodoalkane reagents were recommended, in particular acrylamide [7]. *N*-ethylmaleimide [25, 36] and 4-vinylpyridine also provide alternatives [25]. A study by Suttapitugsakul et al. comparing samples alkylated by iodoacetamide, acrylamide, *N*-ethylmaleimide, or 4-vinylpyridine concluded that iodacetamide and acrylamide were most effective based on monitoring levels of free cysteine, *S*-alkylated cysteine and off target alkylation at the peptide N-terminus and the side chains of lysine and aspartic acid. Acrylamide and iodoacetate performed best in terms of highest completion rate of the cysteine alkylation and lowest number of off-target reaction [25, 37]. Similar numbers of proteins were identified (1700–1800) for each alkylation reagent except for *N*-ethylmaleimide, attributed to off-target alkylation of N-termini as the major alkylation event, occurring at higher levels than the target *S*-alkylation. 4-Vinylpyridine was the least effective reagent at *S*-alkylation. The recommendation to use acrylamide as the alkylation reagent of choice is consistent between these studies [7, 25].

Chloroacetamide has been proposed as an iodoacetamide alternative for analysis of ubiquitin linkage [38]. Direct comparison confirms that chloroacetamide has minimal off-target effects relative to iodoacetamide [7, 24, 39], including lack of the lysine dialkylation event that is isobaric to diglycine [7, 24]. A caveat is that chloroacetamide results in up to 40% methionine oxidation, and to a lesser extent tryptophan (mono and di-alkylation), at higher levels than iodoacetamide, which should be noted to differ from a previous study [7]. While these results may reflect different sample preparation and buffers used for reduction, alkylation, and proteolytic digestion, it does signal caution in the use of chloroacetamide.

In general, the data lead to the conclusion that reduction and alkylation protocols require i) evaluation on an experiment-by-experiment basis to determine the extent of *S*-alkylation and presence of side reactions leading to off-target alkylation and ii) optimization of the protocol, as far as practical, such that *S*-alkylation is complete with minimal off-target alkylation (*see* **Note 3**). As noted by Mouchahoir, T. and Schiel "Unfortunately, the conditions that are amenable to effective reduction, alkylation, and digestion are often the same conditions that promote unwanted modifications" [34].

2.3 Reducing Complexity?

The benefits of reduction and alkylation need to outweigh the risks to data quality arising from off-target alkylation events. Protocol optimization should include evaluation of alkylation efficiency and type, with focus on the *S*-alkylation reaction achieving completion with minimal off-target reactions. Monitoring by database searching to assess disulfide reduction, *S*-alkylation of cysteine and off-target alkylation is recommended. Debate was sparked by Müller and Winter [7] over evaluating of artifactual chemical modifications, particularly for iodoacetamide. The publication was highlighted as a Featured Publication in the July 2017 Mascot Newsletter from Matrix Science. The finding that there was less off-target alkylation and higher efficiency of *S*-alkylation for in-solution digests relative to in-gel digests for all alkylation reagents tested [7] (*see* **Note 5**) was discussed further in a blog posted by John Cottrell [40], noting that the in-solution data has little +57 (iodoacetamide, carbamidomethyl adduct) except on Cysteine, but +12 is observed. Thiazolidine corresponds to 12 Da, listed in Unimod as a formaldehyde adduct, intriguingly proposed an additional degradation product of alkylated residues [40].

The debate will continue and inform best practice. The benefits to proteomic workflows will then be a reduction in sample complexity, which is the aim of the reduction and alkylation steps, thus avoiding the greater complexity imposed by off-target alkylation. The use of filter-based methods markedly decreased off-target alkylation by removal of excess reagents post reduction and alkylation (*see* **Notes 5** and **6**) [7, 41]. In the future, the application of simple and rapid FASP-like methods for sample cleanup and in-filter digestion could well prove helpful in minimizing the impact of off-target. Examples include Filter Based Protein [42] and S-TRAP methods [40].

3 Notes

1. It is sometimes possible to proteolytically cleave a native protein into peptides, without reduction and alkylation, and this is recommended where a sample is known to be low or absent in cysteine content. Reduction and alkylation can be omitted or avoided as a method of protein unfolding since reduction and alkylation are not necessarily prerequisite for proteolytic cleavage but will result in lower levels of cysteine-containing peptides since disulfide bonds prevent proteolytic cleavage [6, 7].
2. In some protocols, it can be advantageous to omit the reduction and alkylation step, even if the protein sample contains cysteine this can still perform relatively well for protein identification albeit with lower reproducibility [7]. For example, in immunoprecipitation protocols, employing microgram quantities. Mohammed et al. noted that reducing and alkylating samples for the rapid immunoprecipitation mass spectrometry of endogenous protein (RIME) protocol, using magnetic beads as the affinity matrix, leads to a >8- to 10-fold increase in the number of peptide spectrum matches (PSMs) to peptides

originating from immunoglobulins. The impact is a reduction of >50% in the number of proteins identified. Removing the reduction and alkylation steps improves sensitivity in terms of protein identification, with lower sequence coverage [43].

3. Optimization of *S*-alkylation workflows using iodoacetamide is detailed for individual proteins and complex samples [25, 28, 34]. A key observation is that off target with iodoacetamide can be limited by incubation at the lower temperature of 4 °C relative to room temperature [33, 34]. Choice of denaturant, buffer, reaction time, and concentration of alkylation reagent are also key consideration [25, 28, 33, 34].
4. It is advisable to follow the recommendation to include a quenching step, specifically using a thiol such as DTT. TCEP and related phosphines are chemically distinct from thiols, therefore non-reactive with alkylating reagents. A quenching step has been suggested to be less critical for in-gel digestion since reagents are removed prior to proteolytic digestion [26] (*see* **Note 5**).
5. In-gel digested samples have been demonstrated to result in significantly higher numbers of off-target residues compared with the in-solution digests [7]. This is proposed to reflect more efficient removal of chemicals from in-solution by Filter Assisted Sample Preparation (FASP) in this study, whereas reduction and alkylation agents enter into gel pieces via diffusion, enabling off-targets reactions to proceed for longer, evidenced by the higher degree of modified peptide N termini for in-gel data set which occurs post cleavage by trypsin [7].
6. FASP also efficiently eliminate most cysteine side reactions that are artifacts. These include β-elimination (desulfuration/loss of H_2S/−34 Da) [7] and oxidation to cysteine sulfenic, sulfinic, and sulfonic acid (+15.9949, +31.9898, and + +47.9847 Da, respectively). Bayer and König discuss these and other artifactual modifications of cysteine in detail in relation to analysis of modified peptides [44].

References

1. Wedemeyer WJ, Welker E, Narayan M et al (2000) Disulfide bonds and protein folding. Biochemistry 39:4207–4216
2. Lindahl M, Mata-Cabana A, Kieselbach T (2011) The disulfide proteome and other reactive cysteine proteomes: analysis and functional significance. Antioxid Redox Signal 14:2581–2642
3. Lundell N, Schreitmüller T (1999) Sample preparation for peptide mapping — a pharmaceutical quality-control perspective. Anal Biochem 266:31–47
4. Herbert B, Galvani M, Hamdan M et al (2001) Reduction and alkylation of proteins in preparation of two-dimensional map analysis: why, when, and how? Electrophoresis 22:2046–2057
5. Padula MP, Berry IJ, Raymond B et al (2017) A comprehensive guide for performing sample preparation and top-down protein analysis. Proteome 5:11

6. Sechi S, Chait BT (1998) Modification of cysteine residues by alkylation. A tool in peptide mapping and protein identification. Anal Chem 70:5150–5158
7. Müller T, Winter D (2017) Systematic evaluation of protein reduction and alkylation reveals massive unspecific side effects by iodine-containing reagents. Mol Cell Proteomics 6:1173–1187
8. Wojdyla K, Rogowska-Wrzesinska A (2015) Differential alkylation-based redox proteomics–lessons learnt. Redox Biol 6:240–252
9. Dzieciatkowska M, Hill R, Hansen KC (2014) GeLC-MS/MS analysis of complex protein mixtures. In: Shotgun proteomics. Humana Press, New York, NY, pp 53–66
10. Shevchenko A, Tomas H, Havlis J et al (2006) In-gel digestion for mass spectrometric characterization of proteins and proteomes. Nat Protoc 1:2856–2860
11. Giansanti P, Tsiatsiani L, Low TY et al (2016) Six alternative proteases for mass spectrometry-based proteomics beyond trypsin. Nat Protoc 11:993
12. Bittremieux W, Tabb DL, Impens F et al (2018) Quality control in mass spectrometry-based proteomics. Mass Spectrom Rev 37: 697–711
13. Herbert B, Hopwood F, Oxley D et al (2003) β-elimination: an unexpected artefact in proteome analysis. Proteomics 3:826–831
14. Galvani M, Rovatti L, Hamdan M et al (2001) Protein alkylation in the presence/absence of thiourea in proteome analysis: a matrix assisted laser desorption/ionization-time of flight-mass spectrometry investigation. Electrophoresis 22:2066–2074
15. Liang X, Wang JR, Wong KWV et al (2014) Optimization of 2-dimensional gel electrophoresis for proteomic studies of solid tumor tissue samples. Mol Med Rep 9:626–632
16. Wu X, Xu C, Wang W (2017) Reduction and alkylation of proteins in 2D gel electrophoresis: before or after isoelectric focusing? Front Chem 5:59
17. Mineki R, Taka H, Fujimura T et al (2002) In situ alkylation with acrylamide for identification of cysteinyl residues in proteins during one- and two-dimensional sodium dodecyl sulphate-polyacrylamide gel electrophoresis. Proteomics 2:1672–1681
18. Lane LC (1978) A simple method for stabilizing protein-sulfhydryl groups during SDS-gel electrophoresis. Anal Biochem 86:655–664
19. Turko IV, Sechi S (2007) Acrylamide—a cysteine alkylating reagent for quantitative proteomics. Methods Mol Biol 359:1–16
20. Fischer R, Kessler BM (2015) Gel-aided sample preparation (GASP)—a simplified method for gel-assisted proteomic sample generation from protein extracts and intact cells. Proteomics 15:1224–1229
21. D'Silva AM, Hyett JA, Coorssen JR (2017) A routine "top-down" approach to analysis of the human serum proteome. Proteome 5:13
22. Naryzhny S (2018) Inventory of proteoforms as a current challenge of proteomics: some technical aspects. J Proteome 191:22–28
23. Wang J, Zhao X, Zhao Y et al (2013) Influence of off-target in enzymatic digestion on the qualitative and quantitative analysis of proteins. Se Pu 31:927–933
24. Hains PG, Robinson PJ (2017) The impact of commonly used alkylating agents on artifactual peptide modification. J Proteome Res 16:3443–3447
25. Suttapitugsakul S, Xiao H, Smeekens J et al (2017) Evaluation and optimization of reduction and alkylation methods to maximize peptide identification with MS-based proteomics. Mol BioSyst 13:2574–2582
26. Boja ES, Fales HM (2001) Off-target of a protein digest with iodoacetamide. Anal Chem 73:3576–3582
27. Woods AG, Sokolowska I, Darie CC (2012) Identification of consistent alkylation of cysteine-less peptides in a proteomics experiment. Biochem Biophys Res Commun 419:305–308
28. Rebecchi KR, Go EP, Xu L et al (2011) A general protease digestion procedure for optimal protein sequence coverage and post-translational modifications analysis of recombinant glycoproteins: application to the characterization of human lysyl oxidase-like 2 glycosylation. Anal Chem 83:8484–8491
29. Yang Z, Attygalle AB (2007) LC/MS characterization of undesired products formed during iodoacetamide derivatization of sulfhydryl groups of peptides. J Mass Spectrom 42: 233–243
30. Lapko VN, Smith DL, Smith JB (2000) Identification of an artifact in the mass spectrometry of proteins derivatized with iodoacetamide. J Mass Spectrom 35:572–575
31. Kruger R, Hung CW, Edelson-Averbukh M et al (2005) Iodoacetamide-alkylated methionine can mimic neutral loss of phos- phoric acid from phosphopeptides as exemplified by nano-electrospray ionization quadrupole time-of-flight parent ion scanning. Rapid Commun Mass Spectrom 19:1709–1716
32. Guo M, Weng G, Yin D et al (2015) Identification of the over alkylation sites of a protein by IAM in

MALDI-TOF/TOF tandem mass spectrometry. RSC Adv 5:103662–103668

33. Xu P, Duong DM, Seyfried NT et al (2009) Quantitative proteomics reveals the function of unconventional ubiquitin chains in proteasomal degradation. Cell 137:133–145
34. Mouchahoir T, Schiel JE (2018) Development of an LC-MS/MS peptide mapping protocol for the NISTmAb. Anal Bioanal Chem 410:2111–2126
35. Creasy DM, Cottrell JS (2004) Unimod: protein modifications for mass spectrometry. Proteomics 4:1534–1536
36. Paulech J, Solis N, Cordwell SJ (2013) Characterization of reaction conditions providing rapid and specific cysteine alkylation for peptide-based mass spectrometry. Biochim Biophys Acta 1834:372–379
37. Huang J, Wang J, Li Q et al (2017) Enzyme and chemical assisted N-terminal blocked peptides analysis, ENCHANT, as a selective proteomics approach complementary to conventional shotgun approach. J Proteome Res 17:212–221
38. Nielsen ML, Vermeulen M, Bonaldi T et al (2008) Iodoacetamide-induced artifact mimics ubiquitination in mass spectrometry. Nat Methods 5:459
39. Schnatbaum K, Zolg D, Wenschuh H et al (2016) Fast and accurate determination of cysteine reduction and alkylation efficacy in proteomics workflows. JPT Application Note
40. http://www.matrixscience.com/blog/step-away-from-the-iodoacetamide.html
41. Wiśniewski JR, Zougman A, Nagaraj N et al (2009) Universal sample preparation method for proteome analysis. Nat Methods 6:59
42. Stepanova E, Gygi SP, Paulo JA (2018) Filter-based protein digestion (FPD): a detergent-free and scaffold-based strategy for TMT workflows. J Proteome Res 17:1227–1234
43. Mohammed H, Taylor C, Brown GD et al (2016) Rapid immunoprecipitation mass spectrometry of endogenous proteins (RIME) for analysis of chromatin complexes. Nat Protoc 11:316
44. Bayer M, König S (2016) Abundant cysteine side reactions in traditional buffers interfere with the analysis of posttranslational modifications and protein quantification—how to compromise. Rapid Commun Mass Spectrom 30:1823–1828
45. Ackermann D, König S (2018) Comparative two-dimensional fluorescence gel electrophoresis. In: Difference gel electrophoresis. Humana Press, New York, NY, pp 69–78
46. Herbert BR, Molloy MP, Gooley AA et al (1998) Improved protein solubility in two-dimensional electrophoresis using tributyl phosphine as reducing agent. Electrophoresis 19:845–851
47. Liu P, O'Mara BW, Warrack BM et al (2010) A tris (2-carboxyethyl) phosphine (TCEP) related cleavage on cysteine-containing proteins. J Am Soc Mass Spectrom 21: 837–844
48. Wang Z, Rejtar T, Zhou ZS et al (2010) Desulfurization of cysteine-containing peptides resulting from sample preparation for protein characterization by mass spectrometry. Rapid Commun Mass Spectrom 24:267–275
49. Zwyssig A, Schneider EM, Zeltner M et al (2017) Protein reduction and dialysis free work-up through phosphines immobilized on a magnetic support: TCEP-functionalized carbon-coated cobalt nanoparticles. Chemistry 23:8585–8589
50. Hale JE, Butler JP, Gelfanova V et al (2004) A simplified procedure for the reduction and alkylation of cysteine residues in proteins prior to proteolytic digestion and mass spectral analysis. Anal Biochem 333:174–181
51. Svozil J, Bärenfaller K (2017) A cautionary tale on the inclusion of variable posttranslational modifications in database-dependent searches of mass spectrometry data. Methods Enzymol 586:433–452
52. HaileMariam M, Eguez RV, Singh H et al (2018) S-Trap is an ultrafast sample preparation approach for shotgun proteomics. J Proteome Res 17:2917–2924

Chapter 8

Detection of Unknown Chemical Adduct Modifications on Proteins: From Wet to Dry Laboratory

Paola Antinori, Théo Michelot, Pierre Lescuyer, Markus Müller, and Adelina E. Acosta-Martin

Abstract

The detection and characterization of chemical adducts on proteins is of increasing interest. Here, we described a step-by-step procedure to identify unknown chemical adduct modifications on proteins resulting from the interaction with a given reactive compound. The protocol can be divided into two equally important parts: (1) the wet laboratory work, to produce high quality mass spectrometry (MS) data of in vitro modified proteins and (2) the dry laboratory work, to analyze the generated MS data and provide highly confident qualitative and quantitative results on the chemical composition and amino acid localization of adducts. This protocol is applicable to the study of any pharmaceutical or chemical compound forming covalent protein adducts, detectable in LC-MS/MS experiments.

Key words High-resolution mass spectrometry, Bioinformatics, Open modification search, Protein adduct modification, Dependent peptide search, R script, MaxQuant, Busulfan, Albumin, Hemoglobin

1 Introduction

The detection and characterization of chemical adducts on proteins is of increasing interest in a large range of applications including environmental and xenobiotics exposure monitoring, in vivo dose determination and toxicological evaluation during drug administration, health risk assessment after chemical agent exposure, metabolism study, and diseases diagnosis through the measurement of metabolic adducts [1]. Here, we described a step-by-step procedure to identify unknown chemical adduct modifications on proteins resulting from the interaction with a given reactive compound, from the generation of adducts in the wet laboratory to the structural hypothesis and validation of results

Electronic supplementary material: The online version of this chapter (https://doi.org/10.1007/978-1-4939-9232-4_8) contains supplementary material, which is available to authorized users.

Caroline A. Evans et al. (eds.), *Mass Spectrometry of Proteins: Methods and Protocols*, Methods in Molecular Biology, vol. 1977, https://doi.org/10.1007/978-1-4939-9232-4_8, © Springer Science+Business Media, LLC, part of Springer Nature 2019

with different bioinformatics tools. This protocol is a modified version of the workflow applied in our previous study [2], in which we have replaced some of the bioinformatics steps by the use of freely available and currently supported software to obtain equivalent results. By doing this, the protocol has become more largely applicable and easy to use as many of these bioinformatics tools are already well accepted and stablished by the proteomics community. As proof of concept, we have applied the present protocol to reproduce previous analyses for the identification of chemical adducts generated by a bi-functional alkylating drug: busulfan, on two blood proteins: human serum albumin and hemoglobin. The relevance of the study of busulfan adducts on major blood proteins relies on the use of this drug as a chemotherapeutic agent [3] administered intravenously as a component of conditioning regimens before hematopoietic stem cell transplantation in adults and children [4].

Protocol steps can be divided into two equally important parts: (1) the wet laboratory work to generate high quality mass spectrometry (MS) data and (2) the dry laboratory work to analyze the data and provide highly confident results. In the wet laboratory work, we include: the selection of the chemical and protein of interest as well as appropriate chemical controls; protein incubations with chemicals; filtration to get rid of unbound chemical excess; denaturation, disulfide bond reduction, cysteine alkylation, and tryptic digestion of incubated proteins; desalting of digested peptides; and MS analysis using a high-resolution tandem mass spectrometer. The dry laboratory work comprises: the initial protein identification and open modification (OMS) search of modification mass shifts; integration of OMS results and quantitative comparison of mass shifts in the different incubation conditions; identification of modified amino acid residues; generation of hypothetical molecular formula and structures of chemical adducts; confirmation of OMS results by a second database search including the proposed adduct modifications; and finally, the annotation and visual inspection of tandem mass spectra from modified peptides.

This protocol is applicable to the study of any pharmaceutical or chemical compound forming stable protein adducts. Also, to any other kind of covalent protein modification that can be generated in a sufficient amount to become statistically significant in its quantitative comparison against negative control incubations. We have not applied this protocol to low abundant modifications resulting from enzymatic reactions. We hope that the protocol in this current version helps to characterize protein adducts from reactive chemicals, including pharmaceutical compounds, in any relevant application, such as monitoring of drug exposure, understanding of pharmacokinetics, and decreasing toxic side effects.

2 Materials

All solutions must be prepared with HPLC-MS grade water and solvents.

2.1 In vitro Protein Incubation with Chemicals

1. Sodium hydroxide (NaOH) solution: 5% in water.
2. Purified human serum albumin (HSA) solution: 15.5 pmol/μL in Dulbecco's Phosphate-buffered saline (D-PBS) pH 7.4. pH of D-PBS can be adjusted with 5% NaOH.
3. Purified hemoglobin (Hb) A0 solution: 15.5 pmol/μL in D-PBS pH 7.4. pH of D-PBS can be adjusted with 5% NaOH.
4. Busulfan solution: 10 nmol/μL in dimethyl sulfoxide DMSO.
5. Butyl methanesulfonate (BMS) solution: 10 nmol/μL in DMSO.
6. Sulfolane solution: 10 nmol/μL in 50% methanol.
7. Precision balance to prepare protein and chemical solutions.
8. PCR 100 μL tubes and PCR thermal cycler (set at 37 °C).

2.2 Filtration

1. Amicon Ultra-0.5 mL, MWCO 3 kDa.
2. Refrigerated micro-centrifuge.
3. Speed-vacuum concentrator.

2.3 Denaturation, Reduction, Alkylation, and Tryptic Digestion

Glassware must be rinsed with ultrapure water to ensure a detergent free environment. All solutions need to be freshly prepared.

1. Buffer solution: 0.1 M Trimethylammonium bicarbonate (TEAB) buffer pH 8 (*see* **Note 1**).
2. Denaturing buffer: 6 M Urea in 0.1 M TEAB buffer pH 8, adjust the pH with 25% HCl (*see* **Note 2**).
3. Reducing agent solution: 30 mM Tris(2-carboxyethyl)phosphine (TCEP) in 0.1 M TEAB pH 8.
4. Alkylating agent solution: 130 mM Iodoacetamide (IAA) in 0.1 M TEAB pH 8.
5. Proteolytic enzyme solution: 0.1 μg/μL of sequencing grade modified trypsin in 0.1 M TEAB.
6. Solution to stop enzymatic proteolysis: 10% Formic acid (FA).
7. Thermo-block or oven for overnight digestion (set at 37 °C).
8. Litmus paper (pH 1–14) or pH-meter.

2.4 Desalting with C18 Reverse-Phase Columns

1. C18 macro-spin columns (*see* **Note 3**).
2. Micro-centrifuge.

3. Conditioning and elution solutions: 50% ACN, 0.1% FA in water (v/v).
4. Equilibrating and wash solutions: 2% ACN, 0.1% FA in water (v/v).
5. Speed-vacuum concentrator.

2.5 Reverse-Phase Nano Liquid Chromatography-Mass Spectrometry (nLC-MS)

2.5.1 Sample Solution: 2% ACN, 0.1% FA in Water (v/v)

1. Nano flow HPLC system (*see* **Note 4**).
2. LC mobile phase A: 0.1% FA in water (v/v).
3. LC mobile phase B: 0.1% FA in ACN (v/v).
4. Trap column: Home-made 0.1 ID × 20 mm length fused silica capillary packed with Magic C18 AQ stationary phase (5 μm, 200 Å pore size) or commercial equivalent (*see* **Note 5**).
5. Analytical nano-column: commercial 0.075 ID × 150 mm length, C18, 5 μm, 100 Å pore size Nikkyo column or equivalent.
6. High-resolution mass spectrometer with nano-spray source: LTQ Orbitrap Velos Pro or other high-resolution instrument with the possibility of HCD ion fragmentation (*see* **Note 6**).

2.6 Bioinformatic Analysis

1. Computer: Needs to be able to run the software indicated below. As this might change over time, please visit the specific web pages for more details about PC requirements and software installation.
2. Fasta files for HSA (UniProtKB accession: P02768) and Hb alpha and beta chains (UniProtKB accession: P69905 and P68871) downloaded from Uniprot (http://www.uniprot.org/).
3. MaxQuant software to run the initial and validation protein identification searches (http://www.coxdocs.org/doku.php?id=maxquant:start). Installation of MSFileReader software (Thermofisher) and Net framework 4.5 (Microsoft) is required to use MaxQuant.
4. R and an R editor software such as Rstudio to integrate and quantify open modification search (OMS) results provided by MaxQuant (https://www.r-project.org/ and https://www.rstudio.com/).
5. Molecular Weight Calculator software to hypothesize molecular formula of chemical adducts (https://omics.pnl.gov/software/molecular-weight-calculator).
6. MSConvertGUI software to generate mgf files from RAW files (http://proteowizard.sourceforge.net/downloads.shtml).
7. pLabel software for mass spectra peak labeling and manual inspection of spectrum-modified peptide matches (http://pfind.ict.ac.cn/software/pLabel/index.html).

3 Methods

3.1 Selection of the Chemical/Protein of Interest and Appropriate Chemical Controls

1. The protocol is intended for use with compounds that potentially generate covalent chemical adducts on proteins, i.e., with nucleophilic or electrophilic reactive groups. Non-covalent interactions will not be detected by this protocol. In our case, we are applying the protocol to busulfan, a bi-functional alkylating agent, which is known to react with DNA to form DNA-DNA intra-strand crosslinks.
2. The possibility of generating the potential modification on interest in vitro as the result of a direct chemical interaction (non-enzymatic) is a requirement to apply the protocol successfully. In principle, adducts generated as the results of enzymatic reactions will be in a much lower amount and might be missed during the statistical analysis of MS data.
3. Select control compounds that help in the interpretation of the data. For example, having a simplified chemical structure of the compound of interest or with known reactivity. In our case, we selected BMS, which shares busulfan chemical structure but has just one methanesulfonate group instead of two, and sulfolane, a known non-reactive busulfan metabolite.
4. Reactive chemical solutions should be prepared in solvent ensuring solubility. Include a negative control condition in which only the drug solvent is added to the incubations. We used DMSO, which is the solvent used to prepare busulfan solutions.
5. Select the protein of interest according to the particular question you are addressing. For example, we have selected HSA and Hb, as they are abundant circulating proteins and clinically relevant for the potential detection of busulfan adducts after intra-venous treatment. In order to simulate blood physiological environment, incubation reactions are performed at pH 7.4 and 37 ° C.

3.2 Protein Incubation with Chemicals

All incubations should be performed at least in triplicates to increase statistical significance and allow reliable interpretation of further quantitative results.

1. For each protein (HSA or Hb), mix 50 μL of protein solution with 11.6 μL of each chemical solution (Busulfan, BMS, or sulfolane) in separate 100 μL PCR tubes (1:150 protein-to-chemical agent ratio) (*see* **Note 7**).
2. As an additional control, mix 50 μL of protein solution with 11.6 μL of DMSO in separate 100 μL PCR tubes.

3. Incubate at 37 ° C for 6 h in a PCR thermocycler (*see* **Note 8**). To summarize, there is a total of 24 incubation tubes, corresponding to triplicate incubations of either HSA or Hb with busulfan, BMS, sulfolane, and DMSO.

3.3 Filtration

1. Set the micro-centrifuge temperature to 4 °C.
2. Add 140 μL of HPLC grade water to each sample (sample volume at this stage: 190 μL).
3. Filter reaction solutions with Amicon Ultra-0.5 filters to eliminate busulfan, BMS, and sulfolane from samples at this point (*see* **Note 9**).
4. Wash the filter by adding 500 μL of water and centrifuge for 35 min at 14,000 × *g* and 4 °C (*see* **Note 10**).
5. Discard the flow-through.
6. Charge sample volume and continue according to manufacturer's instructions (*see* **Note 11**).
7. Evaporate filtered samples under speed-vacuum. After evaporation keep protein pellets for further steps. If necessary, storage at −20 °C at this stage is possible.

3.4 Denaturation, Disulfide Bond Reduction, Cysteine Alkylation, and Tryptic Digestion

1. Turn on the thermo-block and set it at 37 °C.
2. Solubilize protein pellets (around 50 μg of protein) with 29 μL of denaturing buffer.
3. Reduce disulfide bonds with 1 μL of reducing agent solution (final concentration 1 mM) and incubate at 37 ° C for 1 h.
4. Alkylate reduced cysteines with 2 μL of alkylating agent solution (final concentration 8 mM) and incubate at RT for 1 h in the dark and shaking.
5. Dilute urea concentration of protein samples to approximately 1.5 M by adding 120 μL of 0.1 M TEAB (*see* **Note 12**).
6. Add 10 μL of enzyme solution (1:50 enzyme-to-protein ratio).
7. Carry out enzymatic digestion at 37 °C overnight.
8. Stop the enzymatic digestion by adding 10 μL of 10% FA to the peptide solution (pH needs to be approximately 2).

3.5 Peptide Desalting/Cleaning with C18 Spin Columns

Peptide desalting on C18 spin columns can be performed following the manufacturers' instructions. Below we describe the in-house protocol that we currently use (*see* **Note 13**):

1. Condition the column by adding 500 μL of conditioning solution, centrifuge at 2000 × *g* for 4 min and discard the flow-through. Repeat this step using 200 μL (*see* **Note 14**).

2. Equilibrate the column by adding 400 μL of equilibration solution, centrifuge at 2000 × *g* for 4 min and discard the flow-through. Repeat this step (*see* **Note 15**).
3. Add peptide sample on the column, centrifuge for 4 min at 2000 × *g*, and discard the flow-through.
4. Wash by adding 400 μL of washing solution, centrifuge at 2000 × *g* for 4 min, and discard the flow-through. Repeat this step (*see* **Note 16**).
5. Place the column into a new tube. Elute desalted peptides by adding 150 μL of elution solution, centrifuge at 2000 × *g* for 4 min. Keep the flow-through and repeat this step twice using 150 μL and 200 μL (*see* **Note 17**). Final volumes of eluted peptides will be around 500 μL.
6. Prepare aliquots of the eluted peptides, i.e., 10 μg aliquots, evaporate to dryness under speed-vacuum, and store at −20 °C until LC-MS analysis.

3.6 nLC-MS Analysis

1. Solubilize dried peptides in sample solution to obtain a concentration of 0.4 μg/μL. Vortex thoroughly.
2. Inject each sample in triplicate. Use an estimated peptide amount of 2 μg for each injection in the HPLC-MS system. Total number of injections is 72 (3 × 24 incubations).
3. The nLC and MS methods are prepared using Xcalibur software (*see* **Note 18**). LC method: analytical separation at a flow rate of 220 nL/min and a gradient 0–1 min at 5% B, then linear increase to 35% B at 55 min, followed by a linear increase to 80% B at 65 min.
4. MS method: data-dependent acquisition with survey scans in positive ion mode acquired in the FT-orbitrap analyzer using a m/z window from 400 to 2000, a resolution of 60,000, and an automatic gain control target setting of 5×10^5. The three most intense precursor ions were selected for the acquisition of tandem mass spectra (7500 resolution) using higher-energy collisional dissociation (HCD) prior to FT-orbitrap detection (*see* **Note 19**). Charge states 2+ and higher were included for precursor selection. Monoisotopic precursor selection was activated, normalized collision energy was set to 30%, activation time to 0.1 ms, isolation width to 2.5 *Th*, and automatic gain control target value was set to 2×10^5.

3.7 Initial Protein Identification and OMS

1. Use HSA and HbA0 fasta files to configure sequence databases as indicated in MaxQuant online instructions.
2. Load all RAW files of the same incubation type into MaxQuant (nine files (triplicate LC-MS analysis of triplicate incubations)

for each of the four incubation types (busulfan, BMS, sulfolane, and DMSO) for each protein (HSA and Hb)).

3. Run protein identification searches for each of the eight incubation groups with the corresponding protein database and default search parameters, excluding contaminant sequences from the Global Parameters tab and enabling "Dependent Peptides" option with default parameters in the Adv. Identification tab. The dependent peptide option allows an OMS of mass shifts for unknown peptide modifications. The identification of such modified peptides depends on the identification or their unmodified counterparts [5].
4. Once the search finishes, MaxQuant generates a folder named "txt" inside the "combined" folder where the RAW files are located. Copy and rename the eight allPeptides.txt files generated for each search into a separate new folder. These files contain information about the mass shifts detected for unknown modifications and will be used to run the R scripts, i.e., columns whose headers start with DP, referring to dependent peptide identifications.

3.8 Integration of OMS Results and Quantitative Analysis of Dependent Peptide Mass Shifts

Use the script "plotZscore.R," written in programming language R and provided in the supplementary material, to integrate and quantify MaxQuant dependent peptide search results of the different incubation conditions (DMSO, busulfan, sulfolane, BMS). This script has been adapted from [2].

1. Open an R editor (e.g., RStudio).
2. Copy the "plotZscore.R" script into the folder where you have copied the allPeptides.txt files of MaxQuant searches. Set this folder as working directory typing the following command line in the Console window: > setwd("/path/to/your/folder").
3. Open the "plotZscore.R" script with RStudio.
4. In the code, change the names of the conditions to be compared and the names of the allPeptides.txt files as the files in your folder (*see* **Note 20**).
5. If necessary, change the range of mass shifts to be considered (between 0 and 160 Da by default). Values are taken from the column "DP Mass Difference" and only DP peptides with "DP PEP" < 0.01 are kept for calculations.
6. Run the script by typing the following command line: > source("plotZscore.R").
7. When the script finishes, in the plot tab, browse the generated plots. There will be two types of plots (*see* **Note 21**): (1) Z-scores against the mass shifts for each comparison (Fig. 1a) and (2) histograms of the clustered mass shifts that appeared to

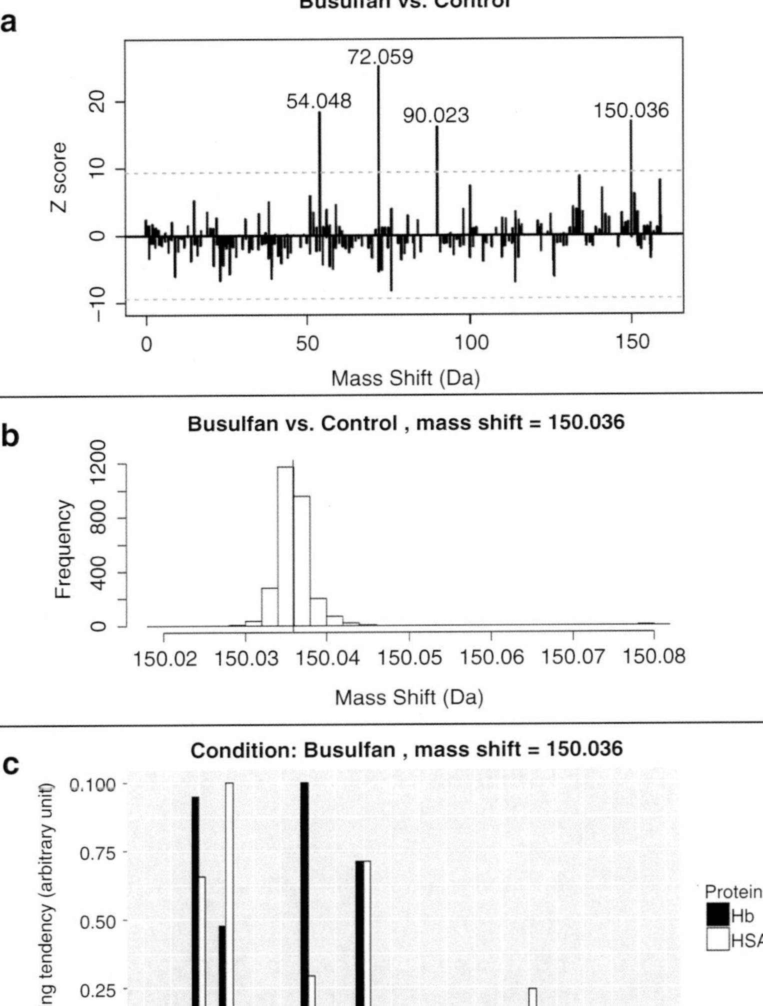

Fig. 1 R script outputs. (**a**) Example of Z-score histograms generated by the plotZscore.R script. Zscore are plotted against mass shifts for the two compared conditions. Z-score of conditions that were defined in "Cond1" in the script code will be represented with positive values and in "Cond2" with negative values. Negative control incubations must be defined in "Cond2" when comparing samples as the Z-score threshold of significance (gray dotted lines) is defined as the absolute value of the most negative Z-score, assuming that the negative Z-scores, enriched in controls (where no adducts should be formed), are noise. In the example, four mass shifts appear to be significant in busulfan incubations compared to controls (54.048, 72.059, 90.023 and 150.036). (**b**) Example of mass shift distributions generated by the plotZ-score.R script. The frequency of mass shifts with significant Z-scores in the different condition comparisons is plotted individually. A representative value of each mass shift is obtained as the median of the values clustered within one bin width (i.e., 0.1) of the center of the corresponding bin. In the example, the distribution and median value of the mass shift 150.036 is shown (from Z-score comparison between busulfan and control incubations). (**c**) Example of positioning tendency histogram generated by the plotPosition.R script. Amino acid positioning tendencies for each protein are calculated as a weighted frequency of "DP AA" values using the "DP positional probability" values in the MaxQuant allPeptides.txt file. These frequencies are normalized by dividing by the largest value to obtain comparable results for the two proteins. In the example, D, E, H, and K are the amino acids of both proteins with highest positioning tendencies for the mass shift 150.036 in busulfan incubations

be significant in the Z-score plots with their representative median value (Fig. 1b). These median values will be used later on to generate hypothetical molecular formula of chemical adducts.

3.9 Amino Acid Positioning of Mass Shifts

Use the script "plotPosition.R," written in programming language R and provided in the supplementary material, to calculate the tendency of a given modification mass shift to be located on a particular amino acid residue. This script has been adapted from [2] to use "DP positional probability" and "DP AA" values of the mass shift of interest.

1. Copy the "plotPosition.R" script into the "R" folder that you have already created.
2. Open the "plotPosition.R" script with RStudio.
3. Install the ggplot2 R package by typing the following command line in the Console window: > install.packages("ggplot2"). This package is necessary for the generation of the positioning plots.
4. In the code, change the names of the condition to be considered, the names of the corresponding data files (one for Hb and one for HSA), and the mass shift to be considered (150.036 Da by default). The script only works for one mass shift of a particular condition at a time.
5. Run the script by typing the following command line: > source ("plotPosition.R").
6. When the script finishes, in the plot tab, normalized positioning tendencies for each amino acid will be represented separately for each protein (Fig. 1c).

3.10 Generation of Hypothetical Molecular Formula and Structures of Chemical Adducts

1. Open the "Formula Finder" tool in Molecular Weight Calculator software.
2. Select all elements composing the chemical of interest, adding custom elements if necessary. Select "Match Molecular Weight" mode and activate "Ppm Mode" and "Show Delta Mass." In "Formula Finder Options," activate "Smart H atoms" and "Sort Results," and select "Sort by Abs(Delta Mass)" and "Thorough Search."
3. Introduce mass shift median values provided by the "plotZscore.R" script into the "Molecular Weight of Target" box. Use a reasonable "Weight Tolerance," i.e., 10 ppm, and press "Calculate" button. The hit at the top of the list (with the lowest absolute mass difference) will correspond to the molecular formula of the modification mass shift, which will be used to configure adduct modifications in MaxQuant later on. If no hits appear with 10 ppm, increase the value until getting a candidate formula.

4. Hypothesize chemical structures of adduct modifications by taking into account the chemical structure of the compound of interest used for incubations, the amino acid residues modified by that chemical, and the molecular formula of the adduct mass shifts.

3.11 Confirmation of OMS Results via Database Searches

1. From the analysis of OMS results on the dependent peptide search, select the adducts of your interest and configure their modifications in MaxQuant software as described in the online instructions: BMS adduct modification (+56.0625968 Da for +C4H9(-H) on Asp, Glu, and His), primary busulfan adduct modification (+150.035063 Da for +C5H11O3S(-H) on Cys, Asp, Glu, His, Lys, and Arg), and busulfan intra-peptide cross-link (busulfan loop; +54.0469476 Da for +C4H8(-2H) on Asp, Glu, and His). These modifications will be added as variable when running protein identification searches.
2. Load all RAW files of the same protein into MaxQuant for busulfan and BMS incubations (nine files each).
3. Run protein identification searches for each of the four incubation groups with the corresponding protein database, variable modifications, and default search parameters.
4. Once the database search finishes, the list of modified peptides with their corresponding mass matched spectrum scan number will be in the "evidence.txt" file inside the "txt" folder generated by MaxQuant.

3.12 Annotation and Visual Inspection of Tandem Mass Spectra

1. Generate mgf from RAW files of busulfan and BMS incubations using MSConvertGUI software (*see* **Note 22**).
2. Using Excel or equivalent software, order the data in each "evidence.txt" file by the following column headers: increasing "Raw file," decreasing "Busulfan (DEH)" (or BMS, depending on the search performed), increasing "Sequence," and increasing "MS/MS Scan number." All busulfan/BMS modified peptides will be at the top of the table, their sequence and MS/MS scan number will be used to browse and annotate MS/MS spectra from mgf files with pLabel software.
3. Open pLabel (a short and very clear user guide is available at the same download internet web site) and click on the "ION" tab to enter the mass of the new amino acid modifications that you want to validate as you have done in the previous step with MaxQuant (busulfan modifications: +54.0469476 Da for +C4H8(-2H) and +150.035063 Da for +C5H11O3S(-H), and BMS modification: +56.0625968 Da for +C4H9(-H) on Asp, Glu, and His), primary busulfan adduct modification (+150.035063 Da for +C5H11O3S(-H) on Cys, Asp, Glu, His, Lys, and Arg), and busulfan intra-peptide cross-link (busulfan loop; on Asp, Glu, and His).

4. Upload the mgf file with the modified peptides that you wish to inspect into the viewer "File" → "Load MS/MSfile."
5. Go to the excel file with the list of modified peptides and use that information to search for the first MS2 scan in pLabel. Go to the "Main info" tab at the bottom left of the page. Paste the peptide sequence of that scan match in the "SEQ" window, the peptide sequence with the annotated y and b ions appears on the up right corner. You can now select the adduct modification from the "MOD" window.
6. In the same tab, select monoisotopic mass measurement, introduce the MS2 mass tolerance (we choose a comfortable 15 ppm tolerance for Orbitrap data), and the intensity threshold depending on the noise signal in your data. Peaks under the threshold will not be labeled (for our HCD MS2 spectra, we set the threshold to 2.5%) (*see* **Note 23**).
7. In the "Ion Type select" tab, choose the singly, doubly, and triply charged fragment ions that you wish to match depending on the ion fragmentation method (CID, ETD, HCD, EThcD).
8. In the low m/z range of the spectra, you can look for signature immonium ions carrying the modification. Moreover, you can increase the sequence coverage of your peptide by assigning peaks corresponding to internal ions (not matched by most search engines). Increasing the sequence coverage allows to assign the position of the modification with higher confidence.
9. As an indication, you can consider that an amino acid residue is indeed modified if three or more consecutive ions of the same series (b or y for CID/HCD data) including the modification site are confidently labeled during the visual inspection.
10. Sometimes the chemical modification of an amino acid can change the fragmentation pattern of the peptide. It is always recommended to compare the mass spectra of the modified and unmodified forms of the peptide at the same charge state.

4 Notes

1. Trypsin activity is optimal at pH 8. The pH of TEAB can be variable, from 8.5 it can go up to 9; to lower it to approximately 8 add HCl 25% starting with 5 μL for 50 mL of buffer, measure the pH and, if necessary, add more.
2. To prepare the denaturing buffer, first weight urea, then add TEAB 1 M, and bring it to volume with water stirring with a magnet bar. When the powder is solubilized the volume of the solution goes down. Take the magnet out and adjust the vol-

ume with water. The denaturing solution with urea and TEAB 0.1 M has a pH higher than TEAB only. Measure the pH and add HCl 25% to lower the pH to 8.

3. Many different desalting devices are available; choose the macro or micro—spin columns depending on the amount of peptides you have to purify. They are compatible with bench micro-centrifuges.
4. Most mass spectrometers for proteomic analysis are hyphenated to nano flow HPLC systems such as Waters NanoAquity or Thermo Scientific Ultimate 3000 RSLCnano.
5. If you wish to pack your own columns here's the link for a useful video http://www.benchfly.com/video/96/how-to-pack-a-capillary-column/. The results might be less reproducible than with a commercial column.
6. The advantage of using HCD fragmentation over other high-resolution fragmentations is that it provides low m/z fragments such as immonium ions, which will be very helpful in data interpretation.
7. Concentrations and protein-to-chemical ratio should be chosen in adequacy to ensure the generation of enough chemical adduct modification to be detected.
8. Since the reactions take place in a small volume, the PCR thermocycler avoids evaporation. In case of larger reaction volumes, regular Eppendorf tubes might be incubated in a thermos-block or oven.
9. Before filtering reaction solutions, filter efficiency needs to be tested with chemical solutions only (before incubation) to check whether they are suitable or not [2]. In the next steps the pH has to be raised from 7.4 to 8 and lowered to 2. Due to these dramatic pH changes and the fact that other chemicals are added to treat the sample, it is necessary to minimize the possibilities for uncontrolled reactions.
10. There's glycerine on the filter and it's important to get rid of it since it interferes with MS analysis.
11. The filter cutoff is 3 kDa and the max sample volume allowed is 500 μL. During filtering, HSA and Hb are confined in the upper part of the filter since their MWs are approximately 65 kDa and 16 kDa each while the small molecules pass through the filter and are discarded. Filters with different cutoffs and upper chamber volumes are available on the market.
12. Urea concentration should be lower than 2 M to avoid trypsin denaturation.
13. It's important to provide a clean peptide mixture to the LC-MS analysis. Mass spectrometers are very sensitive and "non-peptidic" species present in the solution can potentially trigger a

signal for acquisition in the mass spectrometer at the expenses of the time spent to analyze the peptides.

14. At high organic solvent concentration, the C18 chains on the silica column phase are "disentangled" and exposed to the solvent.
15. Lowering the organic content of the environment favors the interactions between the hydrophobic C18 chains and the peptides, which are attracted to the solid phase and stick to it.
16. The small polar molecules are washed away from the column while the peptides remain attached.
17. Highly organic solution provides a friendly environment for the peptides, which are detached from the solid phase and collected in the recovery tube.
18. The protocol described here is for use with Thermo Scientific Orbitrap instruments and related software. It should be adapted for use with other types of high-resolution MS instruments.
19. By using HCD as fragmentation technique, information about the immonium ions at the low *m/z* end of the MS/MS spectra are gathered. Immonium ions provide specific information about the presence of amino acids in the peptide sequence. When there is an interest in amino acid modifications, as in our case, diagnostic ions can be potentially detected in the low *m/z* region of the MS/MS spectra.
20. How conditions are compared is defined by order in the matrices "allCond1" and "allCond2." So, the first condition in one matrix will be compared to the first condition in the other matrix, the second with the second, and so on. If using only one protein for incubations instead of two, e.g., Hb, the script can be used by comment out all HSA related lines of code (21–24, 31–34, 53, 56, 62, 64, 72, 74, 76–78) and uncomment lines 80–82.
21. It might be possible that the Z-score plot appears like it has been squeezed. If this is the case, export the plot as pdf to visualize it with the right dimensions.
22. Multiple software to generate mgf files are available. We suggest MSConvertGUI as it is free and easy to use and download, but other possibilities are Proteome Discoverer (Thermo) or RawConverter (http://fields.scripps.edu/rawconv/) software.
23. One way to establish a reasonable threshold is to look at the ions of the series *y*, *y*-H_2O, *b*, *b*-H_2O, *a*, and *a*-H_2O. Establish the threshold level looking at the appearance/disappearance of the *y*-H_2O, *b*-H_2O, and *a*-H_2O labels.

Acknowledgments

This study was supported by the Swiss Centre for Applied Human Toxicology and the Swiss National Science Foundation (grant No 32003B_143809). The authors would like to thank the Proteomics Core Facility of the University of Geneva for providing full access to the LTQ Orbitrap MS instrument.

None declared conflict of interest.

References

1. Tornqvist M, Fred C, Haglund J et al (2002) Protein adducts: quantitative and qualitative aspects of their formation, analysis and applications. J Chromatogr B Analyt Technol Biomed Life Sci 778:279–308
2. Acosta-Martin AE, Antinori P, Uppugunduri CR et al (2016) Detection of busulfan adducts on proteins. Rapid Commun Mass Spectrom 30:2517–2528
3. Hassan M, Andersson BS (2013) Role of pharmacogenetics in busulfan/cyclophosphamide conditioning therapy prior to hematopoietic stem cell transplantation. Pharmacogenomics 14:75–87
4. Huezo-Diaz P, Uppugunduri CR, Tyagi AK et al (2014) Pharmacogenetic aspects of drug metabolizing enzymes in busulfan based conditioning prior to allogenic hematopoietic stem cell transplantation in children. Curr Drug Metab 15:251–264
5. Tyanova S, Temu T, Cox J (2016) The MaxQuant computational platform for mass spectrometry-based shotgun proteomics. Nat Protoc 11:2301–2319

Chapter 9

Considerations for Identifying Endogenous Protein Complexes from Tissue via Immunoaffinity Purification and Quantitative Mass Spectrometry

Joel D. Federspiel and Ileana M. Cristea

Abstract

Protein complexes perform key roles in nearly all aspects of biology. Identification of the composition of these complexes offers insights into how different cellular processes are carried out. The use of affinity purification coupled to mass spectrometry has become a method of choice for identifying protein-protein interactions, but has been most frequently applied to cell model systems using tagged and overexpressed bait proteins. Although valuable, this approach can create several potential artifacts due to the presence of a tag on a protein and the higher abundance of the protein of interest (bait). The isolation of endogenous proteins using antibodies raised against the proteins of interest instead of an epitope tag offers a means to examine protein interactions in any cellular or animal model system and without the caveats of overexpressed, tagged proteins. Although conceptually simple, the limited use of this approach has been primarily driven by challenges associated with finding adequate antibodies and experimental conditions for effective isolations. In this chapter, we present a protocol for the optimization of lysis conditions, antibody evaluation, affinity purification, and ultimately identification of protein complexes from endogenous immunoaffinity purifications using quantitative mass spectrometry. We also highlight the increased use of targeted mass spectrometry analyses, such as parallel reaction monitoring (PRM) for orthogonal validation of protein isolation and interactions initially identified via data-dependent mass spectrometry analyses.

Key words Immunoaffinity purification, Protein-protein interactions, Mass spectrometry, Tissue, Quantitative MS, PRM, Parallel reaction monitoring, IP-MS

1 Introduction

The formation of protein interactions and multiprotein complexes underlines the ability of proteins to carry out various functions fundamental to life. Defining these protein interactions and how they are regulated is therefore critical for understanding the biology of a system. Immunoaffinity purification coupled to mass spectrometry (IP-MS) has become one of the most frequently implemented methods for identifying protein-protein interactions. These studies have yielded critical biological insights into a variety

Caroline A. Evans et al. (eds.), *Mass Spectrometry of Proteins: Methods and Protocols*, Methods in Molecular Biology, vol. 1977, https://doi.org/10.1007/978-1-4939-9232-4_9, © Springer Science+Business Media, LLC, part of Springer Nature 2019

of important cellular processes, such as transcriptional regulation, intracellular trafficking pathways, immune response, and signaling pathways [1–9]. IP-MS analyses have been most frequently implemented for the isolation of proteins tagged with affinity probes. However, a common limitation of these methods is that these studies typically rely on the overexpression of a tagged version of the protein of interest (herein referred to as the "bait"). This results in the identification of potentially artifactual interactions that are detected because of the nonphysiological abundance levels of the bait, which can lead to cellular stress responses, folding alterations in the tagged protein, and increased nonspecific associations. Additionally, studies of this type are limited to model systems in which introduction of exogenous DNA is possible. Even when genetic engineering methods for a model system exist, such as for mice, the generation of the tagged variant mouse is a time-consuming endeavor.

Immunoaffinity enrichment of the endogenously expressed protein has therefore been of great interest, as this approach can be theoretically applied to any experimental system and removes artifacts associated with tagged-protein overexpression. A caveat for this approach is that it relies on the availability of antibodies that have high affinity and specificity for the bait protein. It is important to note that antibodies validated for another application, such as western blot or immunofluorescence, may still not be suitable for IP. Moreover, the clonality of the antibodies should be taken into account when optimizing the experimental workflow. This can help determine whether one antibody or a mixture of antibodies is needed for access to epitopes (single or multiple) and effective isolation of the bait. Further considerations when performing IP-MS studies from tissues also include the ability to effectively lyse and solubilize samples, which can be impacted by the heterogeneity of tissues when compared to cell culture samples. An optimal lysis buffer condition would provide an efficient solubilization of proteins, while still maintaining endogenous protein-protein interactions. Once the IP conditions have been optimized, quantitative mass spectrometry analyses provide a sensitive and unbiased means to identify interactions and assess their specificity.

Diverse quantitative MS-based proteomic techniques have been paired with IP, including label-free and isotope-labeled methods. In IP experiments from cell culture samples, isotopic labeling with stable isotopes [10] or labeled amino acids in cell culture (SILAC) [11] has been widely used for determining the specificity of interactions [12, 13], relative stability of interactions [1, 2, 14, 15], and even dynamics of interactions in response to perturbations [16]. Although isotopic labeling has proven to be effective for accurate relative quantification of protein interactions, this approach is not easily applicable to all model systems. Although metabolic labeling has been used in animal models, such as mice

[17, 18], this approach can be cost-prohibitive and time consuming. Label-free proteomics, on the other hand, has no additional cost, relies on intrinsic quantitative information in the form of either integrated MS1 peak area or spectral counting, can be incorporated into any proteomic platform, is compatible with any model system, and has been used extensively to investigate protein-protein interactions [4, 5, 19–22]. As such, label-free quantitative MS is readily paired with endogenous protein complex enrichment experiments. Quantification using targeted MS, such as Parallel Reaction Monitoring (PRM), has recently become another valuable tool in protein interaction studies. As detailed in this chapter, targeted MS can be applied to assessing the solubility of proteins, efficiency of bait isolation, and validation of interactions.

Regardless of which MS analysis method is selected, one of the most important considerations in protein interaction studies is the assessment of the specificity of the identified interactions. The increasing sensitivity of mass spectrometry instrumentation provides improved depth of sample analysis, which in practice translates to an increased number of identified specific and nonspecific associations. Several computational algorithms have been developed to assist in assessing interaction specificity when using label-free data [23–27]. These tools are invaluable when performing endogenous IP studies, as the antibodies available for IP are typically not as specific as epitope tag antibodies, and thus have more nonspecificity and cross-reactivity. Furthermore, for endogenous IPs, the usual negative control is a pre-immune IgG IP, which is not as good of a control as using an antibody against an epitope tag in a cell line that expresses the epitope tag alone. These limitations of endogenous IPs make the inclusion of a flexible specificity filtering method even more critical. One of the available tools for specificity filtering is the Significance Analysis of INTeractome (SAINT) [24, 28]. SAINT makes use of label-free quantitative data (either spectral counts or MS1 peak areas) to create a probabilistic mixture model that can compare negative control and bait IPs and generate a specificity assessment for each binary bait-prey interaction. These scores can be used to assign a specificity threshold to filter out most of the nonspecifically interacting proteins.

In this chapter, we present a detailed protocol that takes into account all the abovementioned considerations of IP-MS experiments for the isolation of an endogenous protein from tissue samples (Fig. 1). We begin by discussing the selection of optimal lysis conditions for the bait and the assessment of these conditions by targeted mass spectrometry. We describe how to evaluate antibodies to determine their suitability for IP-MS studies, and how to test solid bead supports for protein isolation. Following the completion of bait isolation, the steps for sample preparation for MS analysis are detailed. Finally, we illustrate how to analyze the resulting list of co-isolated proteins in order to assess the specificity of inter-

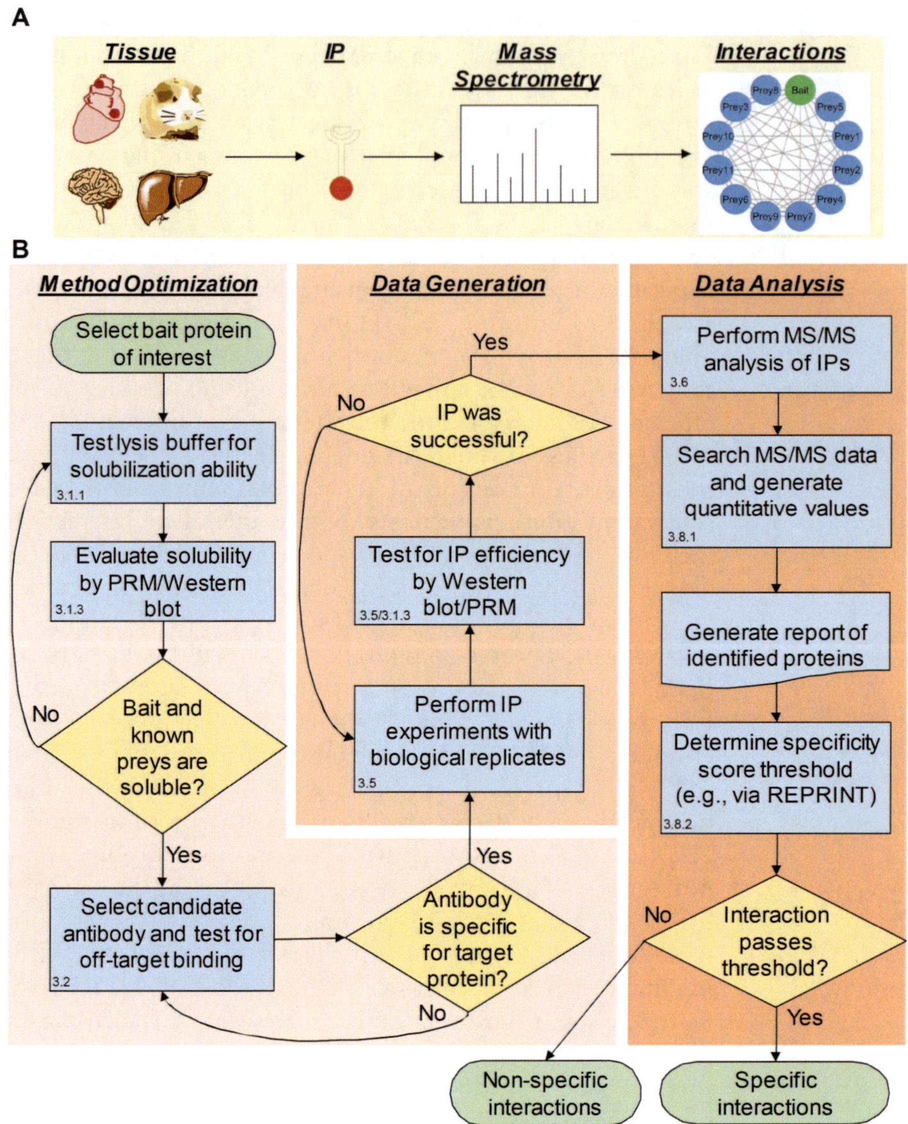

Fig. 1 (**a**) The general goals of an IP-MS experiment are to go from tissue to molecular insights. (**b**) The workflow for taking a target bait protein of interest to a final set of specificity-filtered protein-protein interactions is shown as a flowchart using standard symbols. Decision points where a previous step may need to be repeated for optimization are indicated. The corresponding sections of this chapter are provided on each node for reference

action, as well as the likely functional interaction networks that these proteins form. Throughout the chapter, we highlight the value of using PRM to assess the performance of different steps in the IP-MS workflow (Fig. 2). Where specific branded chemicals or components are listed, these are used routinely in our protocols, alternatives are available.

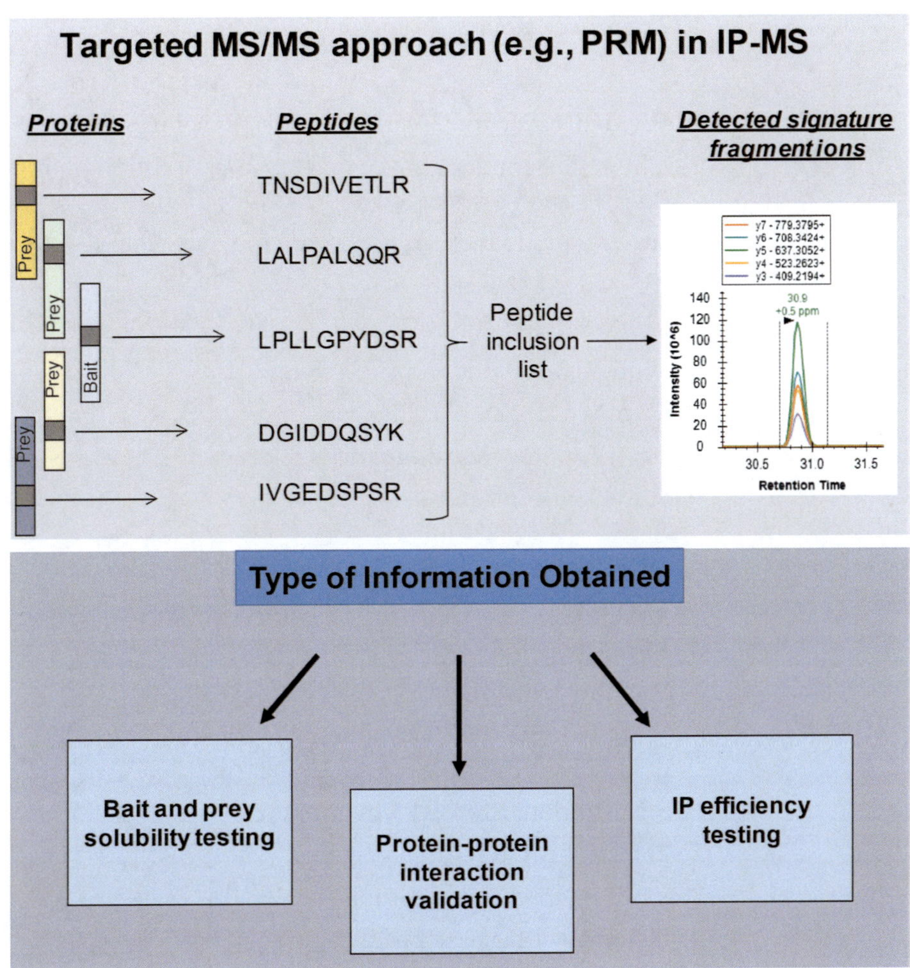

Fig. 2 Applications of PRM in IP-MS studies. PRM has varied uses in IP-MS experiments and can replace the use of western blotting analyses for solubility testing, IP efficiency verification, and validation of identified interacting proteins with an increased throughput and confidence in the results than western blot allows

2 Materials

2.1 Tissue Lysis

2.1.1 Lysis Buffer Selection

1. Detergents: Triton X-100, Igepal CA-630 (NP-40 replacement), Tween-20, Deoxycholate, Digitonin, CHAPS, etc.
2. Salts: NaCl, KCl, $CaCl_2$, $MgSO_4$, KOAc, $MgCl_2$, $ZnCl_2$, etc.
3. Buffers: HEPES, Tris, etc.
4. Protein assay: Bicinchoninic Acid (BCA) or other compatible assay.
5. SDS lysis buffer: 50 mM Tris HCl pH 8.0, 100 mM NaCl, 0.5 mM EDTA, 4% w/v SDS.
6. Cup horn or tip sonicator.

2.1.2 Lysis Method Selection

1. Tenbroeck dounce homogenizer.

2.1.3 In-Solution Digestion for PRM of Bait Protein Isolation

1. Bond-Breaker tris(2-carboxyethyl)phosphine TCEP, 0.5 M.
2. Chloroacetamide (CAM): make a stock solution of 500 mM in ddH_2O and store frozen at −20 °C.
3. LC-MS grade methanol: room temperature and −20 °C.
4. LC-MS grade water.
5. LC-MS grade chloroform.
6. Sequencing-grade trypsin: 0.5 μg/μL.
7. 40 mM HEPES pH 8.2.
8. 1.5 mL microcentrifuge tubes.
9. Cup horn sonicator.

2.1.4 StageTip Cleanup of PRM Samples

1. 10% trifluoroacetic acid (TFA).
2. 1% TFA.
3. Microcentrifuge.
4. Vacuum centrifuge.
5. Elution buffer: 5% ammonium hydroxide (v:v), 15% LC-MS grade H_2O, 80% LC-MS grade acetonitrile.
6. 3 M Empore SDB-RPS membrane.
7. 200 μL LoRetention pipet tips.
8. 2 mL microcentrifuge tube.
9. Hamilton 14-gauge needle.
10. Hamilton syringe plunger.
11. 3 mL Combitip.
12. Autosampler vials.
13. Injection solution: 1% formic acid, 1% acetonitrile, 98% water (all LC-MS grade).

2.2 Validation of Antibodies for IP

1. Commercial or homemade antibodies against the target protein of interest.
2. Pre-immune IgG control.

2.3 Bead Selection

1. Protein A/G magnetic beads (Pierce #88802).

2.4 Immunoaffinity Enrichment of Endogenous Protein Complexes and Sample Preparation

1. Protein A/G magnetic beads.
2. Validated antibody against target protein of interest (bait).
3. ddH_2O.
4. Magnetic stand for microcentrifuge tubes (optionally, magnetic stand or bar magnet for larger volume tubes if high volume IPs are performed).

5. Tube rotator.
6. Round-bottom 2 mL microcentrifuge tubes.
7. 1.5 mL microcentrifuge tubes.
8. Tenbroeck dounce homogenizer.
9. Protein assay: BCA or other compatible assay.
10. 50 mL conical tubes.
11. 0.5 M NH_4OH solution: 0.5 M NH_4OH, 0.5 mM EDTA in water.
12. TEL elution buffer: 106 mM Tris HCl, 141 mM Tris Base, 2% LDS, 0.5 mM EDTA. Recommended to make at least 10 mL and then aliquot and store at −20 °C.
13. 1× TBT buffer: 20 mM K-HEPES pH 7.4, 110 mM KOAc, 2 mM $MgCl_2$, 0.1% Tween-20 (v/v), 1 μM $ZnCl_2$, 1 μM $CaCl_2$.
14. Optimized Lysis buffer: determined in Subheading 2.1.1.
15. Halt Protease and Phosphatase Inhibitor Cocktail (100×).
16. Pierce Universal Nuclease for Cell Lysis or equivalent.
17. Dry bath heat block.
18. Bond-Breaker TCEP, 0.5 M. Can be prepared from alternative supplier.
19. Chloroacetamide (CAM): 500 mM.
20. Cup horn sonicator.
21. MS grade Trypsin.

2.5 Data-Dependent Mass Spectrometry

1. Dionex UltiMate 3000 UHPLC.
2. Q Exactive HF mass spectrometer.
3. Integrafrit column (75 μm ID, 25 cm long).
4. ReproSil-Pur C18 AQ.
5. Stainless steel emitter.
6. Mobile phase A: 0.1% formic acid in LC-MS water.
7. Mobile phase B: 0.1% formic acid, 2.9% LC-MS H_2O, 97% LC-MS acetonitrile (all LC-MS grade).
8. Nanospray Flex Ion source.

2.6 PRM Mass Spectrometry

1. Same equipment as Subheading 2.5.
2. An inclusion list of all the peptides to be analyzed by PRM.

2.7 Data Analysis

1. Multi-core Windows computer with at least 24 GB of RAM and a 1 TB hard drive.
2. Proteome Discoverer v2.2.

3. Skyline (https://skyline.ms) [29].
4. REPRINT (https://reprint-apms.org/).
5. Microsoft Excel.

3 Methods

3.1 Tissue Lysis

One of the most important steps in the isolation of endogenous protein complexes from tissues is the initial lysis step. The lysis conditions must be sufficient for solubilizing the proteins of interest, but should not disrupt the non-covalent interactions between proteins in a complex. As such, great care must be exercised in the selection of the lysis buffer and the mechanical means of tissue homogenization. Suggestions and considerations for both aspects of tissue lysis are given in the next two sections. The careful selection of a lysis condition is often the most time-consuming step in an IP-MS experiment.

3.1.1 Lysis Buffer Selection

Proper selection of an optimal lysis buffer for a bait protein requires testing several conditions and can be evaluated in multiple ways. Varying concentrations of detergents, salts, and buffering agents, as well as the pH, can have a profound impact on the solubility of proteins and protein complexes and also impact the ability to isolate complex components. For example, a buffer with high levels of SDS will certainly solubilize most proteins, but it will also disrupt non-covalent protein-protein interactions. Finding the right balance of lysis buffer components is critical to the success of IP-MS experiments. Because the buffer conditions must be empirically determined for each protein, it is best to initially perform smaller scale pilot lysis experiments to test several different buffer conditions. As a starting point, a nonionic detergent like Triton X-100 or Igepal CA-630 used at concentrations ranging from 0.2 to 1% along with 150 mM NaCl in a HEPES-based buffer system at pH 7.4 will work well for many non-membrane-associated proteins. For nuclear proteins, an increase in the salt concentration (e.g., 250 mM NaCl) may be necessary; if the protein associates with DNA, then treatment with a DNase may be needed to reduce background (this can eliminate real interactions that are DNA-dependent, which should be taken into account). Membrane-associated proteins could require a different detergent like digitonin or CHAPS to be solubilized and the salt concentrations or buffer components may need to be altered. Common detergents used in IP-MS studies with comments about when these detergents should be considered are shown in Table 1.

The impact of the lysis buffer can be tested in multiple ways. Most simply, the presence of the bait protein in the buffer-soluble and buffer-insoluble fractions can be tested. This can be done by

Table 1
Commonly used detergents in IP-MS studies

Detergent	Type	Comments
Triton X-100	Nonionic	Often used at a 0.1–1.0% concentration in lysis buffers. Mild and non-denaturing
Igepal CA-630	Nonionic	Also known as Nonidet P-40. Mild and non-denaturing. Often used at a 0.1–1% concentration. Considered milder than Triton
Tween-20	Nonionic	Mild and non-denaturing. Typically not used as the sole detergent in lysis buffers, but rather present at a low percentage in combination with another detergent
Digitonin	Nonionic	Good for solubilizing lipids and allowing access to lipid-associated proteins. Therefore, a good option for membrane-bound proteins
Deoxycholate	Anionic	Harsher than the nonionic detergents above. Solubilizes lipids. Often added at low percentages <0.5% to lysis buffer in addition to another detergent to increase solubility and decrease nonspecific background interactions
CHAPS	Zwitterionic	Harsher than nonionic detergents. Also good at solubilizing lipids and can be added at low percentages to increase stringency of buffers

western blot analysis or by targeted MS assays like PRM. The advantage of PRM is that one can not only screen for the bait protein but also for other proteins of interest in a single, 1-h analysis to determine how multiple proteins are affected by a lysis condition. For example, if the bait has any known interacting proteins, they can all be monitored at once to determine if a lysis buffer is sufficient for solubilization. Detection by PRM does not guarantee that the lysis buffer maintains protein interactions, but it does ensure that known interacting proteins are in the soluble fraction and thus have a chance to be captured. Furthermore, PRM or western blot analysis can also be used in conjunction with small-scale test IPs to confirm that the bait and known interactions are co-isolated under the lysis conditions used before a full-scale IP effort is undertaken. With these considerations in mind, lysis buffer suitability can be assessed following the steps below.

1. Prepare test buffer in sufficient quantity for the tissue sample, and chill on ice.
2. Solubilize tissue by dounce homogenization as discussed in the next section (Subheading 3.1.2).
3. Centrifuge the lysate at 10,000 × g for 10 min at 4 °C to pellet the insoluble material.
4. Move the soluble fraction (supernatant) to another tube.

5. Dissolve the pellet in SDS buffer and solubilize by repeated rounds (usually two-three rounds) of heating at 95 °C and sonication for 10 × 1 s pulses. This is the buffer-insoluble fraction for future analysis.
6. Perform BCA or other compatible protein assay on buffer-soluble and -insoluble fractions to determine total protein amount in each fraction.
7. Proceed to Subheading 3.1.3 for PRM analysis.

3.1.2 Lysis Method Selection

In addition to selecting an appropriate buffer for solubilization of the proteins of interest, mechanical disruption of the tissue is often necessary to ensure efficient lysis. Previously, we have reported that cryogenic lysis followed by Polytron homogenization was an effective method of cellular disruption for IP when working with samples from cell cultures [14, 30, 31]. However, our experience with mouse brain tissue has been that a traditional dounce homogenization using a Tenbroeck homogenizer provides an effective solubilization across all classes of proteins (*see* **Note 1**).

3.1.3 In-Solution Digestion for PRM Analysis of Bait Protein Isolation

In this section, lysates from above will be prepared for PRM analysis by utilizing a methanol/chloroform precipitation [32] followed by overnight digestion of the protein pellet with trypsin. Unless otherwise stated, all steps should be performed at room temperature.

1. Adjust 50 μg of soluble and insoluble lysates from Subheading 3.1.1 to 1 μg/μL and 2% SDS concentration and add TCEP and CAM to a final concentration of 25 mM and 50 mM, respectively. Note final volume for each sample to be analyzed.
2. Incubate at 70 °C for 20 min.
3. Add 4 × the sample volume of LC-MS methanol (for 50 μL of sample + TCEP + CAM, add 200 μL) and vortex.
4. Add 1 × the sample volume of LC-MS chloroform and vortex.
5. Add 3 × the sample volume of LC-MS water and vortex.
6. Centrifuge at 2000 × *g* for 5 min at room temperature. After this spin, there should be a visible interphase and the top layer should appear clear. If the top layer appears cloudy, spin again.
7. Remove supernatant above the interphase.
8. Add 3 × the sample volume of LC-MS methanol, chilled to −20 °C.
9. Centrifuge at 9000 × *g* for 3 min at 4 °C. The protein will pellet at the bottom of the tube.
10. Remove supernatant without disturbing the pellet.

11. Add 5 × the sample volume of LC-MS grade methanol chilled to −20 °C.
12. Centrifuge at 9000 × *g* for 3 min at 4 °C.
13. Remove all supernatant without disturbing pellet. Pellet should appear wet but no liquid should remain in tube.
14. Add 100 μL of 50 mM HEPES pH 8.2 to the pellet (to achieve a protein concentration of 0.5 μg/μL) and sonicate the tube until the pellet is broken up and the solution appears cloudy (*see* **Note 2**).
15. Add 1 μL of 0.5 μg/μL trypsin to the tube and place in a 37 °C incubator overnight. Proceed to Subheading 3.1.4 the next day.

3.1.4 StageTip Cleanup of PRM Samples

1. Following the digestion in Subheading 3.1.3, add 10% TFA to the samples to a final concentration of 1% and place sample on ice for 5 min.
2. Create one StageTip [33] per sample by cutting two disks out of the SDB-RPS membrane using the 14 gauge needle. Place these disks into a LoRetention pipet tip using the Hamilton syringe plunger to push the disks out of the needle. Pack the disks firmly into the bottom of the pipet tip. Place tip into a 2 mL tube with a hollowed out section in the center of the lid for the tip to go into.
3. Centrifuge acidified samples for 5 min at >10,000 × *g*.
4. Pipet supernatant into the pre-assembled StageTip. Avoid any pellet at the bottom of the tube.
5. Centrifuge StageTip at 2500 × *g* for 2 min to pass all of the peptide solution through the column.
6. Add 200 μL of 1% TFA to the StageTip and spin again for 2 min at 2500 × *g*.
7. Move StageTip to a labeled autosampler vial and add 50 μL of the elution buffer.
8. Use the 3 mL Combitip to manually elute the peptides from the StageTip into the autosampler vial.
9. Place autosampler vial in a vacuum centrifuge and speedvac to almost dryness (<1 μL).
10. Resuspend peptides in injection solution at a concentration of 0.5 μg/μL and proceed to Subheading 3.7.

3.2 Selection and Validation of Antibodies for IP

3.2.1 Selection of IP Antibody

There are several important considerations that must go into antibody selection for IP-MS. The first of these is the availability of antibodies to the target of interest. If no commercial antibody exists, then one must be generated. In the case that multiple antibodies are available (or an antibody is being generated) the clonality of the antibody must be considered. Polyclonal antibodies are generally able to enrich more of the target protein than a monoclonal antibody due to having multiple binding sites (epitopes) on the protein. However, these antibodies can suffer from decreased specificity due to cross-reactivity, and therefore can lead to increased nonspecific background in IP experiments. IPs using monoclonal antibodies are generally cleaner, due to having a single epitope for binding; however, the effectiveness of the isolation depends on the availability of that epitope for binding to the antibody in the context of the IP. If the target epitope is blocked by a protein-protein interaction or protein folding (from either the native state of the protein or a change induced by the lysis conditions), the monoclonal antibody will not be able to bind. This can be partially overcome by using a cocktail of two or more monoclonal antibodies with non-overlapping epitopes, as shown in [34]. This will increase the probability that at least one of the epitopes is accessible for antibody binding, which should allow an increased enrichment of the bait during the IP. A caveat of this approach is that the use of multiple antibodies can lead to an increased disruption of protein-protein interactions. Resources exist to aid in the selection of antibodies for IP, including Antibodypedia [35] through The Human Protein Atlas and product documentation from the manufacturer. Additionally, some antibody companies are beginning to more thoroughly validate their antibodies. For example, ThermoFisher Scientific has recently undertaken an effort to validate many of their antibodies by IP-MS, and vetted antibodies are listed on their website (http://www.thermofisher.com/us/en/home/life-science/antibodies.html) with an advanced verification tag.

3.3 Validation of Selected IP Antibody

Once an efficient lysis method is selected for bait solubilization, and a candidate antibody (or multiple antibodies) chosen, the viability of the antibody for IP in the given model system should be examined. This can be done to a limited extent by IP-western blot analysis to confirm that the antibody is capable of enriching the target protein. While this can be used as a first-pass evaluation, a better test is to perform a pilot IP-MS experiment (following the protocol listed in Subheading 3.4) in the tissue type of interest to ensure that the antibody can indeed enrich the target bait protein and to determine if there are any off-targets [36, 37]. Off-target proteins that are being enriched by the antibody can be difficult to distinguish from real interaction partners of the bait. A potential clue to identify off-targets is if a co-isolated protein is identified at a much higher abundance than the bait protein and is not known as a common contaminant protein. To determine if a protein is a likely IP contaminant that is present due to its intrinsic "stickiness" or high abundance, a negative control database like CRAPome [25] can be consulted. If a protein is not a contaminant but is present at a much higher level in the IP compared to the target bait protein, it is possible that this is an off-target enrichment. Of course, there is also the consideration of stoichiometry, and therefore the possibility that one copy of the bait can interact with multiple copies of another protein. To test this, if there are multiple antibodies available to the target protein, additional IP-MS pilot studies can be performed to determine if this putative off-target is enriched by other antibodies. If not, then this is an indication that this is an off-target, and this antibody is not recommended for IP-MS studies. An example of this is shown in Fig. 3, illustrating the evaluation of several antibodies for their ability to enrich Hdac4 in mouse brain, and the identification of the protein Gja1 as an off-target for the anti-Hdac4 antibody ab12171.

3.4 Bead Selection

An important complement to antibody selection is the selection of the solid support. In general, the bead options available fall into two broad categories: agarose-based supports and magnetic-based supports. Both classes have been used successfully, but the magnetic beads are generally the preferred choice for IP-MS studies. Magnetic beads are generally smaller and thus have lower background binding, and higher capture capacity (by weight). Additionally, the use of the magnet during the wash steps speeds up sample processing, which in turn leads to lower background. Importantly, the use of such surface binding resins does not restrict or interfere with the size of the isolated protein complex, offering a means to isolate larger macromolecular complexes.

For some time, the leading choice in magnetic beads was the Dynabead products manufactured by Dynal/Life Technologies (now a division of Thermo Fisher), as these beads are relatively small (~3 μm) and uniform in shape, and could covalently bind ~4 μg amounts of antibody per milligram of beads [31]. These beads were used with great success in many studies, as they resulted in clean IPs that could be performed with small amounts of antibodies and beads. Recently, however, the ability to couple antibody to Epoxy-activated Dynabeads has decreased. Currently, approximately 0.3 μg of antibody can be conjugated to 1 mg of beads, which is a less than a tenth of what we previously observed. Others have also noted the recent poor performance of these beads [38].

As a result, we tested several other types of beads, and good results were obtained with the Pierce protein A/G magnetic beads. These beads are ~1 μm in diameter, coated with protein AG for non-covalent association with mouse and rabbit antibodies, associate quickly with a magnet, and have a low background. When 10 μg of antibody is incubated with these beads we observed ~7 μg of antibody bound per 25 μL of bead slurry as shown in Fig. 4a. We typically preincubate the antibody with the beads in order to reduce the IP time and maintain a low background. One hour of preincubation is optimal for binding the antibody to the beads as is shown in **step 6** in the protocol provided in Subheading 3.5 below. Depending on the abundance of the bait protein, the amount of beads and corresponding antibody can be scaled by adding additional amounts of both components.

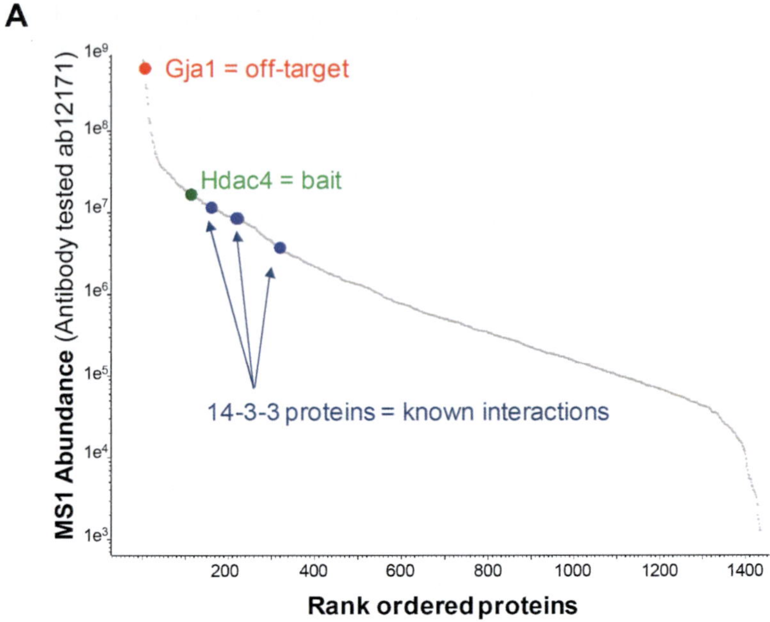

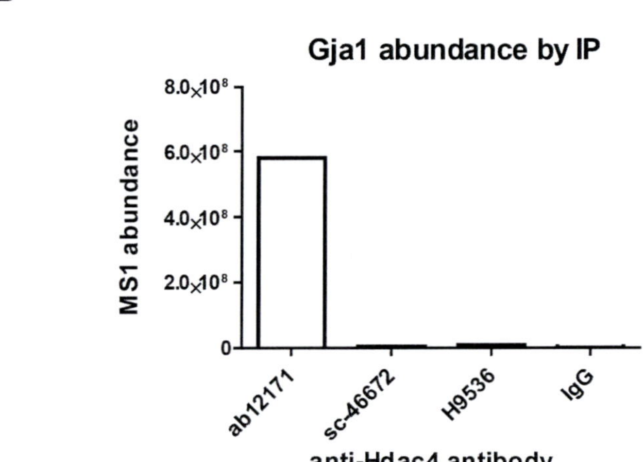

Fig. 3 Assessment of antibody performance in IP-MS studies and the presence of antibody off-targets. (**a**) The MS1-based abundance of identified proteins is graphed from high to low. The target of the antibody (Hdac4, green), a known set of Hdac4 interacting proteins (14-3-3 proteins, blue) and a prominent off-target protein (Gja1, red) are shown in a rank ordered list of protein abundances in an anti-Hac4 IP using antibody A. (**b**) A comparison of several different anti-Hdac4 antibodies demonstrates that Gja1 is an off-target enrichment by antibody ab12171 in mouse brain

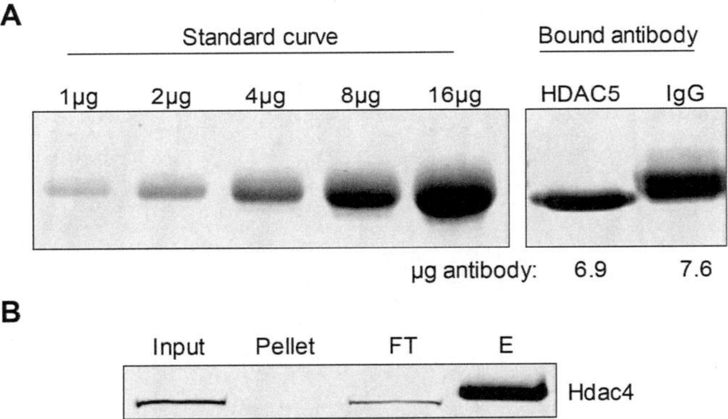

Fig. 4 Pre-conjugation of antibody to protein A/G beads and IP efficiency assessment. (**a**) Antibodies were preincubated with magnetic protein A/G beads for 1 h at 4 °C. By comparing the amount of antibody bound to the beads to the standard curve, approximately 7 μg of antibody can be bound to a 25 μL slurry of beads. (**b**) An example western blot is shown that demonstrates that the bait protein (Hdac4) is solubilized (compare Input and Pellet lanes), depleted from the soluble fraction by the IP (compare flow-through (FT) with Input lanes), and eluted from the beads (lane E)

3.5 Immunoaffinity Enrichment of Endogenous Protein Complexes

During several of the following **steps** (**12**, **14**, **18**, **29**), aliquots of the input, pellet, flow-through, elution 1, and elution 2 will be collected. These samples should be loaded onto an SDS-PAGE gel in equal percentages (e.g., 1% each) to check for efficiency of solubilization, isolation, and elution of the bait protein by western blot. *See* Fig. 4b for an example of this with HDAC4 in mouse brain. As an alternative, aliquots of the input, pellet, and flow-through samples can be processed by in-solution digestion (as in Subheading 3.1.3) and analyzed by PRM for bait levels and any known prey proteins.

1. Prepare a sufficient volume of 1× TBT buffer for coupling all antibodies needed for the IP (2.5 mL per 25 μL slurry).
2. Resuspend protein A/G magnetic bead slurry completely by pipetting up and down. Aliquot 25 μL of the bead slurry for each IP to be performed into a round-bottom 2 mL tube (*see* **Note 3**). Important: Do not vortex the beads at any point until the elution step.
3. Wash beads 3 × 500 μL with 1× TBT buffer using a magnetic stand to collect beads after each wash.
4. Resuspend beads in 500 μL 1× TBT buffer.
5. Add 10 μg of antibody per 25 μL of bead slurry.
6. Incubate the beads for 1 h at 4 °C with rotation (*see* **Note 4**).
7. While the antibody is conjugating to the beads, prepare a sufficient volume of the optimal lysis buffer identified in Subheading 3.1.1 so that the lysate will be at a 1 mg/mL concentration for the IP. Add protease and phosphatase inhibitors

(and optionally, a nuclease) at 1× concentration immediately prior to IP. Also prepare 7 mL of wash buffer (same composition as lysis buffer but lacking inhibitors) per IP. Place all buffers on wet ice along with a dounce homogenizer to prechill prior to IP.

8. Approximately 15 min prior to the end of the antibody conjugation time, begin preparing the tissue lysate. Thaw the tissue on ice for 1–2 min and then place tissue in dounce homogenizer. Add an appropriate volume of lysis buffer and dounce tissue with slow and even strokes (*see* **Note 5**).
9. Transfer the lysate to a prechilled 50 mL conical tube and incubate lysate on ice for 5 min with occasional agitation.
10. Centrifuge the lysate at 10,000 × *g* at 4 °C for 10 min.
11. By this point, the 1 h conjugation of the antibody to the protein A/G beads should have completed. While the lysate is centrifuging, place the beads on a magnetic stand and remove the supernatant. Wash the beads 2 × 1 mL with the wash buffer from step 3.5.7 by gentle pipetting. Following the final wash, resuspend the beads in 100 μL of wash buffer per IP and place on ice until needed.
12. Once the lysate finishes centrifuging, carefully pipette the supernatant into a new prechilled tube. Keep the insoluble tissue pellet for later processing as it will serve as the insoluble fraction for western blot analysis.
13. Perform BCA or other compatible protein assay on cell lysate supernatant to determine protein amount in lysate and adjust to 1 mg/mL concentration if necessary (*see* **Note 6**).
14. Keep an aliquot of the cell lysate supernatant following concentration adjustment to serve as the input fraction for western analysis. Recommended volumes: add 12.5 μL of 4× LDS sample buffer and 2.5 μL of 1 M DTT to 35 μL of input fraction. Use the remaining tissue lysate supernatant for the IP (*see* **Note 7**).
15. Mix the washed antibody beads by gently pipetting up and down. Pipette 100 μL of washed beads into a tube containing the clarified lysate.
16. Rotate the bead-containing lysate on an end-over-end rotator at 4 °C for 1 h.
17. During the incubation, take the insoluble tissue pellet and solubilize in a volume of SDS buffer equal to that of the buffer lysis used for the tissue lysis. Sonicate and heat this pellet to ensure complete solubilization. This will serve as the insoluble fraction for western blot analysis. Recommended volumes: add 12.5 μL of 4× LDS sample buffer and 2.5 μL of 1 M DTT to 35 μL of the insoluble fraction.

18. After the 1 h incubation, use a magnet to collect the antibody beads on one side of the tube. Retain an aliquot of the flow-through (unbound) fraction by transferring the supernatant into a clean tube for further analysis. Recommended volumes: add 12.5 μL of 4× LDS sample buffer and 2.5 μL of 1 M DTT to 35 μL of flow-through. Discard remainder of the flow-through.
19. Resuspend the beads in 1 mL of wash buffer and transfer to a clean 2 mL round-bottom tube.
20. Wash the beads 3 × 1 mL with wash buffer.
21. Discard the last wash buffer wash and resuspend the beads in 1 mL of H_2O chilled to 4 °C. Transfer the beads to a new 2 mL round-bottom tube.
22. Wash beads once more with 1 mL of chilled H_2O and completely remove all water from the tube.
23. Add 200 μL aqueous 0.5 M NH_4OH solution to beads (*see* **Note 8**).
24. Agitate beads for 10 min at room temperature with an automatic mixer or vortex to keep the beads in suspension.
25. Isolate beads on the magnetic rack and transfer the NH_4OH eluate to a microcentrifuge tube. Retain the bead fraction.
26. Repeat **steps 23–25** once more and add the second 200 μL elution to the first (*see* **Note 9**). This combined set of NH_4OH eluates is elution 1.
27. Snap freeze the NH_4OH eluate in liquid nitrogen, then either lyophilize or dry completely in a vacuum centrifuge. This will take several hours.
28. Resuspend the bead fraction in 85 μL 1× TEL buffer, then incubate the beads at 95 °C for 5 min and agitate at room temperature for 10 min. Isolate the eluate and move to a new tube. This is elution 2.
29. Once the NH_4OH elution has dried, resuspend it in 96 μL of 1× TEL buffer. Add 1 μL of 0.5 M TCEP and 3 μL of 0.5 M CAM to both elution 1 and elution 2. Incubate at 70 °C for 20 min to reduce and alkylate cysteine residues. Save 10 μL of each elution for western analysis. The samples can be frozen at this point or immediately processed by proceeding to Subheading 3.6.
30. The elution 1 samples from the IP that were reduced and alkylated in **step 29** are cleaned up and digested in this step using filter-aided sample prep (FASP), as described in a previous Methods in Molecular Biology protocol [14]. Briefly, the IP elution is washed on a molecular weight cutoff filter with a tris-urea-deoxycholate buffer to remove the LDS prior to trypsin

digestion. Then trypsin is added to the filter and the proteins are proteolytically digested overnight. The following day, the peptides are centrifuged through the filter and an ethyl acetate extraction is performed to remove the deoxycholate prior to peptide cleanup. *See* [14] for a detailed protocol. Additionally, 250 ng of trypsin is recommended for digestion of IP samples in this buffer, as this provides a balance between the number of missed cleavages and nonspecific cleavages.

31. The digested peptides from **step 30** are further cleaned up and prepared for MS analysis in this step. The detailed protocol for this procedure can be found in [14]. Briefly, the peptides are passed over a self-packed SDB-RPS StageTip column, washed, and eluted into an autosampler vial. The elution is dried down in a vacuum centrifuge and resuspended in the injection solution. The two recommended changes from the previous protocol are to (a) skip the first two elution steps and proceed directly to elution 3 after the wash steps and (b) resuspend the sample in 5 μL of injection solution. Peptide level fractionation of an IP is generally not necessary when the sample is analyzed on a Q Exactive HF due to the acquisition speed of the instrument. However, if the sample complexity is high (>>1000 proteins), fractionation can be performed as originally described.

3.6 Data-Dependent Mass Spectrometry

1. Check calibration of Q Exactive HF and Ultimate 3000 LC prior to running and recalibrate as needed.
2. The Integrafrit column should be packed with the Reprosil Pur AQ C18 resin to a length of 25 cm and equilibrated by running several standard samples prior to analysis.
3. For IP samples, a single 90 min LC gradient is recommended per sample. The peptides should be separated over a linear gradient from 3 to 30% mobile phase B at a flow rate of 250 nL/min (*see* **Note 10**).
4. A data-dependent tandem MS (ddMS2) experiment should be configured in the Xcalibur method control software. The MS1 scan should be configured as follows: resolution of 120,000, AGC target of 3×10^6, scan range of 380–1800, max inject time of 20 ms. The MS/MS settings should be configured as follows: resolution of 15,000, max inject time of 100 ms, AGC target of 1×10^5, isolation window of 1.2 Da, fixed first mass of 125, loop count of 18, NCE of 27, and an intensity threshold of 1×10^5. These settings should result in a cycle time of less than 2.5 s and allow for sufficient trapping time to identify relatively low abundant peptides.
5. For each sample, 2 μL of the peptides should be injected. This leaves sufficient sample for one extra injection in case of instrumentation issues.

6. Standard samples should be injected before and after IP samples to monitor instrument performance.
7. Once all samples have been analyzed, proceed to step 3.8.1.

3.7 Parallel Reaction Monitoring Mass Spectrometry for Analysis of Bait and Prey Solubilization and Isolation Efficiency

1. Generate an inclusion list of proteotypic peptides (unique, fully tryptic peptides) utilizing the Skyline software (*see* **Note 11**). It is advisable to include three to five peptides per protein that come from different regions of the protein to ensure observed trends are real and not simply artifacts of digestion efficiency or posttranslational modifications.
2. Export a standard inclusion list for the Q Exactive from Skyline and copy the resulting .csv file to the instrument computer.
3. The instrument should be calibrated as in Subheading 3.6.
4. For PRM analysis, a 60 min LC separation time is sufficient for most applications. The peptides should be separated over a linear gradient from 3 to 30% mobile phase B at a flow rate of 250 nL/min (*see* **Note 10**).
5. A PRM experiment should be configured in the Xcalibur method control software as follows: resolution of 30,000, AGC target of 2×10^5, max inject time of 80 ms, isolation window of 0.8, and a fixed first mass of 125 (*see* **Note 12**). Import the inclusion list of peptides to target generated in Subheading 3.7.1 above.
6. Inject 2 μL of peptides (1 μg of peptides on column) per sample.
7. Standard samples should be injected before and after PRM samples to monitor instrument performance.
8. Once all samples have been analyzed, proceed to Subheading 3.8.3.

3.8 Data Analysis

Many different data analysis platforms, both commercial and freeware, can be used to analyze MS/MS data. Here we have described the use of the commercial software Proteome Discoverer 2.2 (Thermo Fisher Scientific) for data-dependent acquisition (DDA) (*see* **Note 13**) and the freely available Skyline software for PRM.

3.8.1 Peptide Identification and Label-Free Protein Quantitation of DDA Data

1. Create a new study in Proteome Discoverer 2.2 and load the instrument-generated data files (.RAW format).
2. Conduct a new analysis and perform the processing step (database search) and consensus step (result file assembly). Proteome Discoverer 2.2 includes default workflows for the Q Exactive which perform identification and MS1 label-free quantitation—these workflows are a good place to start. Ensure that the settings listed below are included (*see* **Note 14**).

3. For the processing step, include the Spectrum Files RC node and the Minora Feature Detector node to perform offline mass accuracy recalibration and feature detection, respectively. Choose a FASTA file appropriate for the origin of the tissue analyzed by IP-MS. Search the DDA data using Sequest HT with settings for a fully tryptic search, precursor mass tolerance of 5 ppm, fragment mass tolerance of 0.02 Da, static carbamidomethylation of cysteine, dynamic oxidation of methionine, dynamic deamidation of asparagine, dynamic loss of methione plus acetylation of the protein N-terminus, and, optionally, dynamic phosphorylation of serine, threonine, and tyrosine. Complete the processing node with the Percolator node to perform PSM validation, and optionally the ptmRS node for PTM site localization assignment.
4. For the consensus step, the Feature Mapper and Precursor Ions Quantifier nodes need to be included in the workflow to match peptides across runs and perform accurate MS1-based quantification. The Feature Mapper node helps to reduce missing values in data-dependent MS experiments by matching MS1 signals for a peptide between different runs, even if that peptide was not selected for fragmentation in every run. These MS1 signals can then be quantified by either taking the max peak intensity or integrating the area under the curve for the peptide. This is an advance over spectral count-based quantitation, which utilizes the number of sequencing events as the quantitative parameter, and is prone to missing values for lower abundant proteins. The Precursor Ions Quantifier node has several settings that may need to be altered depending on the IP data in hand. In general, for IP data that will be analyzed using REPRINT (*see* Subheading 3.8.2) and have ≥3 replicates and negative controls, it is best to leave normalization and scaling off, to use replicate-based resampling for the imputation mode, and to perform pairwise ratio based calculations. Two additional important settings in the consensus workflow are to (a) set the merge mode of the MSF file node to "per file and search engine type" and (b) ensure that the target false discovery rate (FDR) (Strict) settings in the Peptide Validator node are set to 0.01. These settings preserve the spectral count data from each sample and set the FDR to 1%, respectively.
5. Once the processing and consensus workflows have completed, a .pdresult file is generated and can be opened and viewed within Proteome Discoverer 2.2. The data can be filtered to the desired FDR of 1% and to have a minimum number of unique peptides (as well as many additional settings) in order to arrive at a list of highly confident protein identifications. The columns displayed can be adjusted to include only the

desired information. Minimally, for every protein this should include the accession number, protein description, and abundance values for each sample. Additionally, the total number of spectral counts can be viewed for each sample if desired.

6. The dataset should then be exported to an Excel file for further processing using REPRINT.

3.8.2 Interaction Specificity Assessment Using REPRINT

1. Go to https://reprint-apms.org/ to perform a specificity assessment of the IP-MS data. Registration for a free account is recommended for access to all features of the resource.
2. The Excel table generated in Subheading 3.8.1 should be reformatted according to the instructions on the REPRINT website. Either the MS1- or the spectral count-based quantitation can be utilized for REPRINT analysis. The choice of which to use depends on the method of data acquisition. For a Q Exactive HF instrument, MS1-based quantitation will be better as this instrument is typically configured so that the same peptide is not sampled multiple times. Thus, the spectral count data will not be as robust in this setting. For an ion trap instrument that does sample the same peptide multiple times, spectral count-based quantitation may be preferred.
3. Click on Analyze and then Start to get to the upload data screen. Select the organism from which the tissue originated, the type of IP, and the quantitation type. Upload the properly formatted data file.
4. Click Next and select archival controls if desired (*see* **Note 15**).
5. Click Next again and advance to the Score Interactions screen. Here a normalization method for MS1 data can be chosen and settings for different scoring models can be selected. Under the Probabilistic SAINT Score section, select SAINTexpress and run the analysis (*see* **Note 16**). Once the analysis has finished, click on View Results at the bottom right of the screen.
6. The view results screen has several tabs in it with different tools and functionalities. On the result matrix tab, a spreadsheet of the results can be downloaded by clicking on the list hyperlink. Under the result visualizations tab, several graphs can be viewed to evaluate the data and decide on a score threshold to use based on known interacting proteins and how these interacting proteins scored in the SAINT algorithm. These scores are labeled as SP and range from 0 to 1 with 1 being the most specific for each binary set of interactions.
7. There are also two different heat map tabs that can be used to evaluate the data with different filtering thresholds which can be adjusted on the fly. There is also a network tab that can be used to automatically build a protein-protein interaction

network. This network view can be downloaded and imported directly into Cytoscape [39, 40] for additional manipulation.

8. The exact choice of filter threshold will depend on the IP and can vary from bait to bait but, in general, a SAINT score of greater than 0.8 is a good starting point for most datasets. The suite of visualizations available in the REPRINT website can greatly aid in selecting the optimal threshold by making the visualization of known interactions immediately available. Additionally, more traditional receiver operating characteristic curves can also be generated [1]. If no interactions are known for the bait, the histogram of scores can be used to examine the data and identify a good cutoff point that preserves high scoring interactions while removing most of the nonspecific interactions.

3.8.3 Peptide Quantitation of PRM Data

1. The data produced in Subheading 3.7 should be imported into Skyline using the .skyfile that was generated to create the inclusion list. The transition settings should be altered to match the settings used for acquisition (30,000 resolution, etc.). Import the results as single injection replicates in files.
2. The imported results are automatically integrated by Skyline but a manual review is recommended. For each file, the automatic peak picking algorithm should be checked to ensure that the peak was correctly selected. This can be confirmed by comparing the spectrum from the apex of the integrated peptide peak to a confidently identified library spectrum of the same peptide (*see* **Note 17**).
3. Once a peptide has been confidently identified in the PRM experiment, the integration boundaries can be manually adjusted to fix errors introduced by the automated algorithm.
4. The data can then be visualized directly in Skyline or exported to a spreadsheet and imported into a preferred graphing software for further analysis.
5. In the case of utilizing PRM to determine solubilization efficiency, the goal is to have all the peptides from the protein(s) of interest be detected in the soluble fraction and not the insoluble fraction. For an IP efficiency check, the goal would be to detect a significant decrease in all the peptides from the protein(s) of interest in the flow-through sample compared to the input sample.

3.9 Validation and Interpretation of IP-MS Data

Once a set of specificity-filtered interacting proteins have been identified, they are often depicted in the form of an interaction network. One of the most popular tools for network visualization is Cytoscape, and REPRINT has the ability to generate a filtered network that can be imported directly into this program. Careful thought should be given to what the message of the interaction network will be (*see* **Note 18**). As show in Fig. 5, interaction networks can be designed to convey different information. A network can illustrate the similarities and differences in interactions of baits from the same family of proteins, as shown for histone deacetylases HDAC4 and HDAC5 in Fig. 5a. If interactions of a bait under two different conditions are to be compared, one can use a color gradient or shapes to depict the differences. Figure 5b provides an example from the IP of endogenous IFI16, a human antiviral protein, from uninfected cells and cells infected with herpes simplex virus type 1 [41]. Such visualization methods can draw attention to dynamic changes in protein associations following a perturbation. Figure 5c provides an example of how the visualization of interactions can be expanded to include comparisons of multiple conditions. Illustrated are selected IFI16 interactions from uninfected cells compared to three different infection states. As seen in these examples, a network can convey different messages depending on the study presented. As a final step in an IP-MS study, the identified interactions of interest that pass the specificity filter (e.g., SAINT) should be validation via an orthogonal experimental approach. A typical method of validation is to perform a reciprocal isolation, in which the identified prey of interest becomes the new bait. The reciprocal co-isolation of the original bait is usually evaluated by western blot. While this is a valid technique, it is limited in throughput and is typically only used to confirm that several of the identified prey proteins can interact with the original bait protein. The use of PRM for detection of proteins of interests provides an opportunity to expand this validation to include an entire set of SAINT-filtered interacting proteins.

4 Notes

1. A drawback of the dounce homogenization is that the reproducibility of the lysis can vary from one experiment due to user dependency. Furthermore, the disruption of cell structure is occurring in lysis buffer, which allows for more spurious protein-protein associations to form. Cryogenic lysis provides a more reproducible disruption, can improve access to the bait, and can help minimize nonspecific associations (given the disruption of cytoskeletal networks). Additionally, cryogenically lysed material can be stored frozen for long periods of time (months to 1 year), while the dounce-lysed material must be used immediately. However, cryogenic disruption cannot be used when it is desirable to maintain larger structures intact, such as organelles or postsynaptic densities.

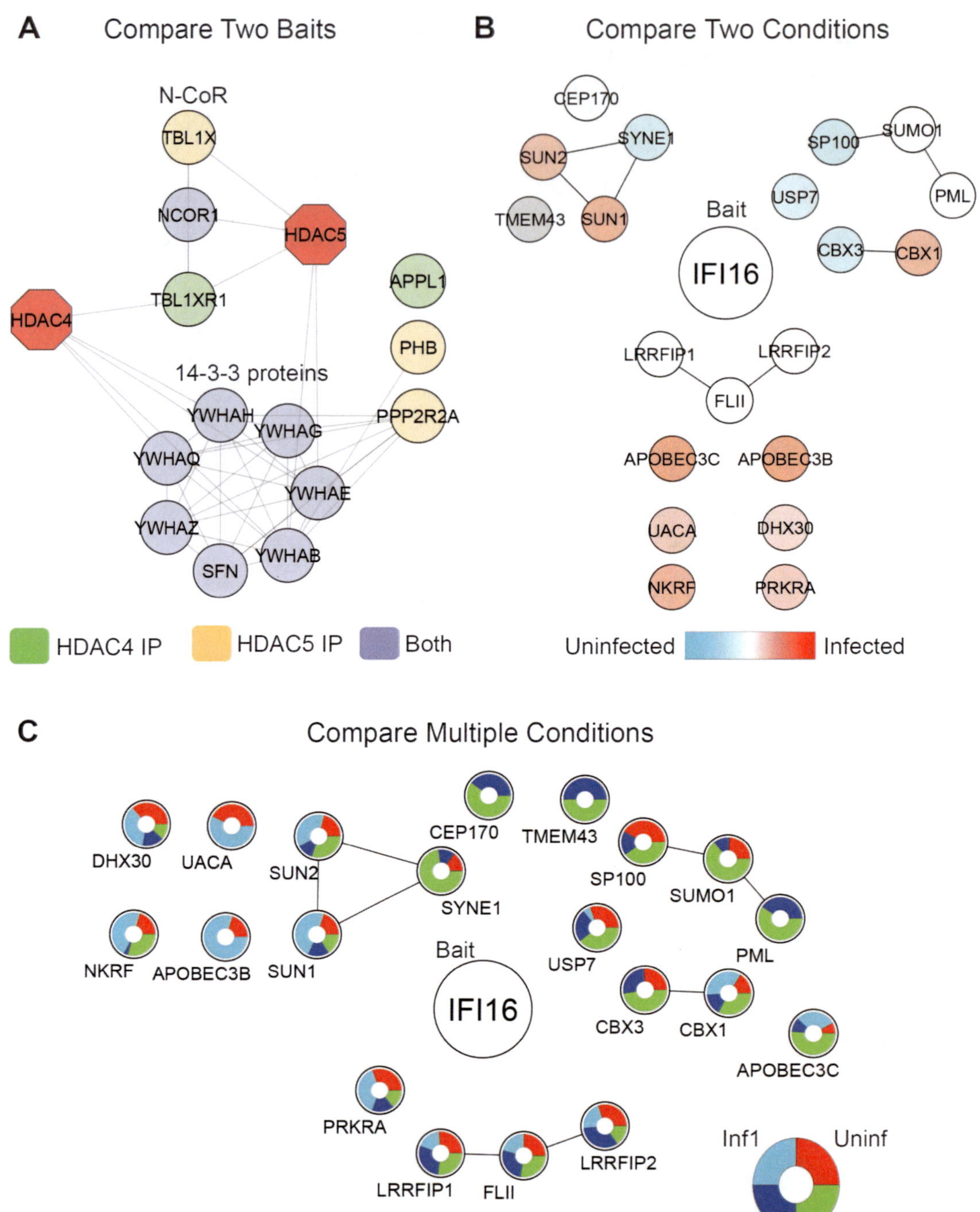

Fig. 5 Examples of interaction networks that compare bait isolations or different biological conditions. (**a**) Two HDAC baits (HDAC4 and HDAC5) from [1] are shown along with specificity-filtered interacting proteins. (**b**) A subset of proteins that interact with IFI16 from [41] are depicted here with relative prey association in either uninfected or HSV-1 wild-type virus infected cells. (**c**) Comparison of interactions in four different biological conditions. The example is shown for the same subset of IFI16 interacting proteins from (**b**). The four compared conditions are: uninfected cells (Uninf), cells infected with wild-type HSV-1 strain (Inf 1), and cells infected with an HSV-1 strain containing a ring finger mutation in the viral ICP0 protein at either 3 h (Inf 2) or 8 h post infection (Inf 3)

2. If the protein pellet proves difficult to resolubilize in HEPES buffer, dilute sample to 0.25 μg/μL and repeat sonication.
3. If performing replicates or other additional IPs with the same antibody, the total antibody amount needed can be conjugated in one tube by scaling up all volumes accordingly.
4. Pre-conjugation of the antibody to the A/G beads prior to IP is recommended to reduce the IP time (and therefore background). However, it is possible that some antibodies may not work as well in this format—if this is the case, then a more traditional sequential addition of the antibody to the lysate for 1 h followed by the addition of the protein AG beads to the lysate for 1 h can be performed.
5. The volume of lysis buffer to be added to the dounce homogenizer depends on the size of the dounce used (which also depends on the size of the tissue to be lysed). For a whole mouse brain (~400 mg of tissue), we use a 7 mL dounce with 5 mL of lysis buffer. The protein concentration during the douncing step can be higher than the 1 mg/mL concentration used for the IP as this can be diluted immediately following dounce homogenization. The number of dounce strokes needed also depends on the tissue. For mouse brain, we recommend 20 strokes as a starting point. Additionally, for the first several strokes it is important to fully dissociate the tissue; we recommend a grinding motion with the pestle during the first few downstrokes to fully break up the tissue. Mincing the tissue with scissors or a scalpel can also assist in breaking up the tissue and ensuring efficient lysis.
6. Performing a protein assay may not be necessary every time an IP is done. If the amount of protein solubilized per amount of tissue is known (likely determined in Subheading 3.1), then a protein assay is not necessary and this previously determined ratio (e.g., protein amount to tissue weight) can be used.
7. It is often effective to lyse a sufficient amount of tissue for two IPs at once and use half for the IP of the target bait protein and the other half for a control IP (IgG, etc.).
8. The aqueous 0.5 M NH_4OH should be made immediately prior to the elution step as it will not maintain the correct pH (10.5) for long periods of time.
9. Check carefully to ensure that no beads are present in the elution tube. It may be necessary to place the tube containing the eluate onto the magnet for 1–2 min and then transfer the eluate to another tube to be certain that no beads are carried forward.
10. If the column heater in the UltiMate 3000 LC is being utilized (recommended), then the equilibration, loading, and washing phases of the LC run can be done at 500 nL/min to reduce run time.

11. There are many excellent tutorials and webinars on the Skyline website that cover all aspects of utilizing this software to design and analyze targeted mass spectrometry experiments. New users are directed to this resource to gain a basic understanding of the software.
12. If more than 20–25 peptides are desired to be targeted in one run, then a scheduled method is advised. This can be done by programming the Q Exactive to only look for a certain peptide during a limited time window based off of prior knowledge of the elution profile of the peptide. In this way, several hundred peptides can be targeted in one 60 min run. Additionally, if it is anticipated that many of the targeted peptides will be highly abundant and a faster MS/MS acquisition time is desired, the maximum injection time can be lowered to 20–50 ms and the resolution to 15,000.
13. While Proteome Discoverer and the algorithms described in this protocol are paid software, several groups have developed freely available analysis nodes that have much of the same functionality as the default workflows. At the time of writing, these nodes can be installed in the Proteome Discoverer 2.1 demo to use it as "freeware."
14. Proteome Discoverer also includes an extensive user manual that can be accessed by clicking Help and User Guide. New users are referred here for a more comprehensive guide for how to customize and configure the search settings.
15. REPRINT has archival negative control data for several organisms as part of the CRAPome repository. However, these are all spectral count-based IP datasets and will not be compatible if intensity-based quantification is used. SAINT analysis can still be performed even if archival controls are not utilized. In this case, it is critical to perform matched control experiments alongside the bait IP.
16. In addition to SAINTexpress, the original SAINT algorithm is available for use on the REPRINT website. This original SAINT algorithm allows for tuning of the scoring parameters which may be advantageous for datasets that have control IPs with significantly lower complexity than the bait IPs. *See* [24] for more information about how to choose the parameters to use with this version of the algorithm.
17. Ideally, a reference spectrum would have been generated on the same instrument. A library spectrum can be generated most easily by identifying the peptide of interest in a data-dependent run. However, if a protein is of low abundance in the model system used, this may not be a feasible option. Additional methods include searching a PRM run with a database search algorithm like Sequest to try to get a confident

automated annotation of a peptide spectrum match, synthesizing a standard peptide and analyzing it by MS/MS, or by overexpressing the protein of interest and using the overexpression system to confidently identify a peptide that can then be used in an endogenous expression system.

18. Given the complexity of protein interactions networks, sometimes such interaction figures lose the ability to provide clear information and have been referred to as "hairballs." However, interaction networks can be valuable and informative if careful thought is put in the creation of these type of figures. Numerous examples of informative networks are available in the literature, some of which include an interactome of all human HDACs [1], a network view of the relationships between protein complexes from more than one quarter of all human proteins [5, 42], and a study of the protein interactions involved in the Hippo pathway [43].

Acknowledgments

We are grateful for funding from the CHDI foundation and from the NIH (GM114141 and HL126509). We thank Todd M. Greco and Elizabeth A. Rowland for helpful comments in the preparation of this manuscript.

References

1. Joshi P, Greco TM, Guise AJ et al (2013) The functional interactome landscape of the human histone deacetylase family. Mol Syst Biol 9:672
2. Budayeva HG, Cristea IM (2016) Human Sirtuin 2 localization, transient interactions, and impact on the proteome point to its role in intracellular trafficking. Mol Cell Proteomics 15(10):3107–3125
3. Diner BA, Li T, Greco TM et al (2015) The functional interactome of PYHIN immune regulators reveals IFIX is a sensor of viral DNA. Mol Syst Biol 11(1):787
4. Kohli P, Bartram MP, Habbig S et al (2014) Label-free quantitative proteomic analysis of the YAP/TAZ interactome. Am J Physiol Cell Physiol 306(9):C805–C818
5. Huttlin EL, Ting L, Bruckner RJ et al (2015) The BioPlex network: a systematic exploration of the human interactome. Cell 162(2):425–440
6. Hubner NC, Bird AW, Cox J et al (2010) Quantitative proteomics combined with BAC TransgeneOmics reveals in vivo protein interactions. J Cell Biol 189(4):739–754
7. Li X, Tran KM, Aziz KE et al (2016) Defining the protein-protein interaction network of the human protein tyrosine phosphatase family. Mol Cell Proteomics 15(9):3030–3044
8. Scifo E, Szwajda A, Soliymani R et al (2015) Quantitative analysis of PPT1 interactome in human neuroblastoma cells. Data Brief 4:207–216
9. Yadav L, Tamene F, Goos H et al (2017) Systematic analysis of human protein phosphatase interactions and dynamics. Cell Syst 4(4):430–444 e5
10. Oda Y, Huang K, Cross FR et al (1999) Accurate quantitation of protein expression and site-specific phosphorylation. Proc Natl Acad Sci U S A 96(12):6591–6596
11. Ong SE, Blagoev B, Kratchmarova I et al (2002) Stable isotope labeling by amino acids in cell culture, SILAC, as a simple and accurate approach to expression proteomics. Mol Cell Proteomics 1(5):376–386
12. Tackett AJ, DeGrasse JA, Sekedat MD et al (2005) I-DIRT, a general method for distinguishing between specific and nonspecific pro-

tein interactions. J Proteome Res 4(5):1752–1756
13. Wang X, Huang L (2008) Identifying dynamic interactors of protein complexes by quantitative mass spectrometry. Mol Cell Proteomics 7(1):46–57
14. Greco TM, Guise AJ, Cristea IM (2016) Determining the composition and stability of protein complexes using an integrated label-free and stable isotope labeling strategy. Methods Mol Biol 1410:39–63
15. Wang X, Huang L (2014) Defining dynamic protein interactions using SILAC-based quantitative mass spectrometry. Methods Mol Biol 1188:191–205
16. Federspiel JD, Codreanu SG, Palubinsky AM et al (2016) Assembly dynamics and stoichiometry of the apoptosis signal-regulating kinase (ASK) signalosome in response to electrophile stress. Mol Cell Proteomics 15(6):1947–1961
17. Toyama BH, Savas JN, Park SK et al (2013) Identification of long-lived proteins reveals exceptional stability of essential cellular structures. Cell 154(5):971–982
18. Zanivan S, Krueger M, Mann M (2012) In vivo quantitative proteomics: the SILAC mouse. Methods Mol Biol 757:435–450
19. Kennedy L, Kaltenbrun E, Greco TM et al (2017) Formation of a TBX20-CASZ1 protein complex is protective against dilated cardiomyopathy and critical for cardiac homeostasis. PLoS Genet 13(9):e1007011
20. Crow MS, Cristea IM (2017) Human antiviral protein IFIX suppresses viral gene expression during herpes simplex virus 1 (HSV-1) infection and is counteracted by virus-induced proteasomal degradation. Mol Cell Proteomics 16(4 suppl 1):S200–s214
21. Diner BA, Lum KK, Toettcher JE et al (2016) Viral DNA sensors IFI16 and cyclic GMP-AMP synthase possess distinct functions in regulating viral gene expression, immune defenses, and apoptotic responses during Herpesvirus infection. MBio 7(6):e01553-16
22. Jager S, Cimermancic P, Gulbahce N et al (2011) Global landscape of HIV-human protein complexes. Nature 481(7381):365–370
23. Goldfarb D, Hast BE, Wang W et al (2014) Spotlite: web application and augmented algorithms for predicting co-complexed proteins from affinity purification—mass spectrometry data. J Proteome Res 13(12):5944–5955
24. Choi H, Larsen B, Lin ZY et al (2011) SAINT: probabilistic scoring of affinity purification—mass spectrometry data. Nat Methods 8(1):70–73
25. Mellacheruvu D, Wright Z, Couzens AL et al (2013) The CRAPome: a contaminant repository for affinity purification-mass spectrometry data. Nat Methods 10(8):730–736
26. Sowa ME, Bennett EJ, Gygi SP et al (2009) Defining the human deubiquitinating enzyme interaction landscape. Cell 138(2):389–403
27. Armean IM, Lilley KS, Trotter MW (2013) Popular computational methods to assess multiprotein complexes derived from label-free affinity purification and mass spectrometry (AP-MS) experiments. Mol Cell Proteomics 12(1):1–13
28. Teo G, Liu G, Zhang J et al (2014) SAINTexpress: improvements and additional features in Significance Analysis of INTeractome software. J Proteome 100:37–43
29. MacLean B, Tomazela DM, Shulman N et al (2010) Skyline: an open source document editor for creating and analyzing targeted proteomics experiments. Bioinformatics 26(7):966–968
30. Cristea IM, Williams R, Chait BT et al (2005) Fluorescent proteins as proteomic probes. Mol Cell Proteomics 4(12):1933–1941
31. Conlon FL, Miteva Y, Kaltenbrun E et al (2012) Immunoisolation of protein complexes from Xenopus. Methods Mol Biol 917: 369–390
32. Wessel D, Flugge UI (1984) A method for the quantitative recovery of protein in dilute solution in the presence of detergents and lipids. Anal Biochem 138(1):141–143
33. Ishihama Y, Rappsilber J, Mann M (2006) Modular stop and go extraction tips with stacked disks for parallel and multidimensional peptide fractionation in proteomics. J Proteome Res 5(4):988–994
34. Li T, Chen J, Cristea IM (2013) Human cytomegalovirus tegument protein pUL83 inhibits IFI16-mediated DNA sensing for immune evasion. Cell Host Microbe 14(5):591–599
35. Alm T, von Feilitzen K, Lundberg E et al (2014) A chromosome-centric analysis of antibodies directed toward the human proteome using Antibodypedia. J Proteome Res 13(3):1669–1676
36. Persson H, Preger C, Marcon E et al (2017) Antibody validation by immunoprecipitation followed by mass spectrometry analysis. Methods Mol Biol 1575:175–187
37. Uhlen M, Bandrowski A, Carr S et al (2016) A proposal for validation of antibodies. Nat Methods 13(10):823–827
38. Mali S, Moree WJ, Mitchell M et al (2016) Observations on different resin strategies for affinity purification mass spectrometry of a tagged protein. Anal Biochem 515:26–32
39. Shannon P, Markiel A, Ozier O et al (2003) Cytoscape: a software environment for inte-

grated models of biomolecular interaction networks. Genome Res 13(11):2498–2504

40. Cline MS, Smoot M, Cerami E et al (2007) Integration of biological networks and gene expression data using Cytoscape. Nat Protoc 2(10):2366–2382
41. Diner BA, Lum KK, Javitt A et al (2015) Interactions of the antiviral factor interferon gamma-inducible protein 16 (IFI16) mediate immune signaling and herpes simplex virus-1 immunosuppression. Mol Cell Proteomics 14(9):2341–2356
42. Huttlin EL, Bruckner RJ, Paulo JA et al (2017) Architecture of the human interactome defines protein communities and disease networks. Nature 545(7655):505–509
43. Couzens AL, Knight JD, Kean MJ et al (2013) Protein interaction network of the mammalian Hippo pathway reveals mechanisms of kinase-phosphatase interactions. Sci Signal 6(302):rs15

Chapter 10

Metaproteomics of Freshwater Microbial Communities

David A. Russo, Narciso Couto, Andrew P. Beckerman, and Jagroop Pandhal

Abstract

Recent advances in metaproteomics have provided us a link between genomic expression and functional characterization of environmental microbial communities. Therefore, the large-scale identification of proteins expressed by environmental microbiomes allows an unprecedented view of their in situ metabolism and function. However, one of the main challenges in metaproteomics remains the lack of robust analytical pipelines. This is especially true for aquatic environments with low protein concentrations and the presence of compounds that are known to interfere with traditional sample preparation pipelines and downstream LC-MS/MS analyses. In this chapter, a semiquantitative method that spans from sample preparation to functional annotation is provided. This method has been shown to provide in-depth and representative results of both the eukaryotic and prokaryotic fractions of freshwater microbiomes.

Key words Algae, Bioinformatics, Environmental sample, Functional analysis, Mass spectrometry, Metaproteomics, Microbial community, Protein extraction, Shotgun proteomics

1 Introduction

Determining the best workflow for a metaproteomic analysis is one of the biggest challenges limiting the development of the field. The absence of standardized methods for protein extraction, purification, quantification, and processing is, in part, due to the extreme diversity of sampling environments. The proteins have to be extracted from a complex matrix while simultaneously guaranteeing a high extraction yield, sufficient purity, and a non-biased extraction procedure [1]. For example, the determination of extraction efficiencies in complex marine biofilms has revealed efficiencies comprised between 0.85% and 15.15% [2]. Therefore, the optimization of the extraction procedure for each set of environmental samples is a prerequisite of a successful metaproteomic study. In this chapter, the objectives are to provide a straightforward method for protein identification from dilute freshwater microbial samples.

Caroline A. Evans et al. (eds.), *Mass Spectrometry of Proteins: Methods and Protocols*, Methods in Molecular Biology, vol. 1977,
https://doi.org/10.1007/978-1-4939-9232-4_10,

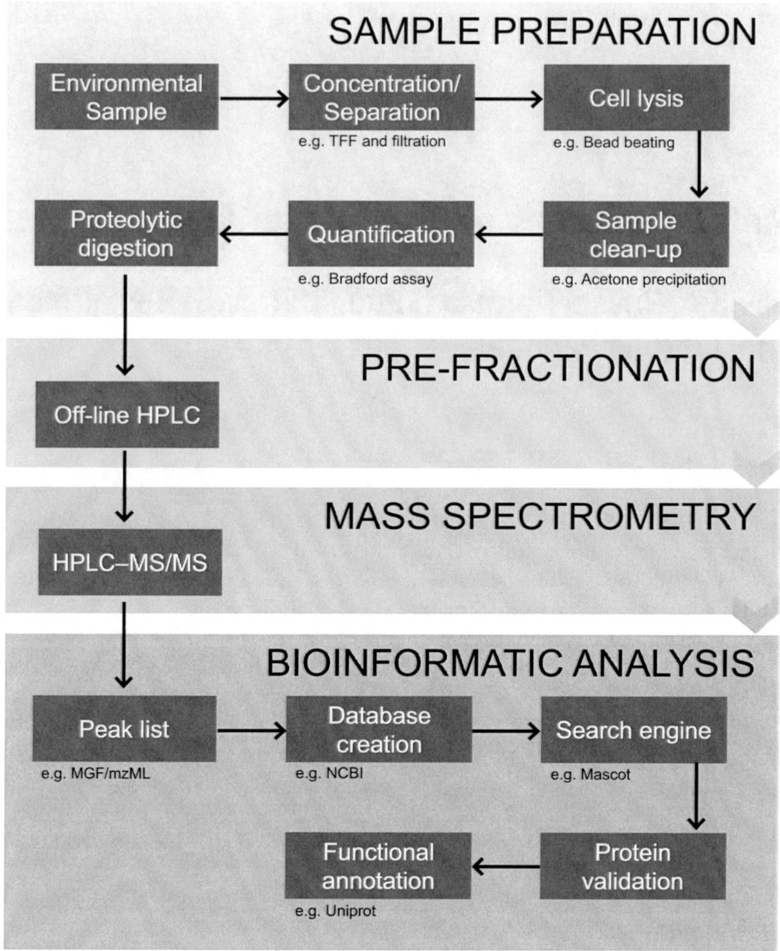

Fig. 1 Example of a metaproteomic workflow designed to maximize protein identification in freshwater microbial environmental samples

The workflow starts with sample collection (Fig. 1). Due to high environmental variability, sampling should be comprehensive to guarantee sufficient representation of the study environment. Unless the target ecosystem is eutrophic, there will typically be a low cell number, and therefore sampling large quantities of water is necessary. The samples are then reduced into a volume appropriate for lab processing, through tangential flow filtration, and subjected to physical fractionation to enrich the samples with individual components of the ecosystem. For example, Teeling et al. [3] sampled 500 L of seawater twice a week. Samples were subsequently filtered into fractions dominated by free-living bacteria (3 to 0.2 μm in size) and algae/particle-associated bacteria (10 to 3 μm in size). This reduced the complexity of the initial samples and increased the representation of lower abundance proteins.

After collection and filtration, several steps are taken to maximize protein extraction yields with an unbiased, reproducible, and mass spectrometry (MS)-compatible method. Protocols for cell disruption vary depending on what type of cells are present in the sample. Physical methods (e.g. bead beating) are sometimes preferred due to superior efficiency and the absence of chemicals which would need to be removed later in the workflow. Extraction buffers are mainly used to control pH but can also include protease inhibitors and nucleases (DNA and RNA degradation). Before further analysis, the extracted protein mixture can be cleaned of any compounds which might negatively affect the enzymatic digestion, fractionation, and/or MS analysis. This can be achieved either by precipitation with trichloroacetic acid, acetone, or ethanol. This step is now followed by gel-based separation or in-solution digestion.

For many years the "gold standard" of protein separation was considered to be 2-D polyacrylamide gel electrophoresis (2-D PAGE). However, this technique is laborious and suffers from several weaknesses (e.g. co-migration of proteins) [4]. More recently, in-solution digestion followed by multidimensional LC-MS has become the experimental standard [5]. An in-solution digestion protocol is not gel dependent and the reduction, alkylation, and digestion are undertaken in the liquid sample. Disulfide bond reduction and alkylation of the cysteine residues can prevent the free sulfhydryls from reforming disulfide bonds. This is followed by an enzymatic protein digestion which will produce a complex peptide mixture that is ready for fractionation and MS analysis.

To improve resolving power, many studies use multidimensional protein identification technology (MudPIT) [5]. This approach utilizes two- or multidimensional chromatography to separate complex peptide mixtures generated by trypsin digestion and is increasingly used for high-throughput analysis of complex environmental samples. A typical two-dimensional approach employs offline liquid chromatography (LC) followed by online LC coupled to MS [6]. After the reduction of sample complexity, the peptides are ionized and analyzed by MS. The most frequently used ionization techniques are matrix assisted laser desorption/ionization (MALDI) and electrospray ionization (ESI). In MALDI, a matrix of embedded peptides is triggered by a laser beam and in ESI a liquid peptide mixture is dispersed by electrospray [7, 8]. These ion mixtures are then analyzed by MS. The most common MS techniques are time of flight (TOF) and ion trap. To achieve high mass accuracy and high resolution hybrid systems such as tandem MS are recommended (e.g., quadrupole-time of flight) [9].

Several bioinformatics tools exist to "decode" the output spectra from the MS. In a first instance, algorithms for protein identification correlate experimental fragment ion spectra with theoretical spectra for each peptide derived after in silico digest of several protein sequences [10]. The most common algorithms used are incorporated in commercial platforms (e.g., MASCOT) [11]; however, several free and open-source alternatives are also available (e.g., X!Tandem, Trans-Proteomic Pipeline) [12]. An aspect that is key to the success of these algorithms is the choice of an appropriate search database. In traditional single-organism proteomics, it is relatively straightforward to develop a detailed database. However, the increasing complexity of the systems targeted by metaproteomic studies creates a unique challenge. Public databases, such as NCBIprot, are incomplete in cases of freshwater microbial samples and often lack proper annotation [13]. Furthermore, non-targeted databases can lead to large search spaces which can inflate false discovery rates. To overcome these issues, identifying the organisms populating a microbial community through amplicon sequencing (e.g., 16S rDNA surveys) or metagenomics shotgun sequencing and utilizing this information to create targeted databases can vastly increase search resolution [14]. Targeted databases can also be coupled with ad-hoc search approaches, such as the two-step approach [15], which are aimed at improving identifications and false discovery rates.

After selecting appropriate databases and search parameters, the search algorithms produce a list of proteins and taxonomic assignments which then have to be validated statistically and quantified. Some platforms incorporate statistical validation and quantification (e.g., MASCOT) or the analysis can be done separately (e.g., Trans-Proteomic Pipeline). Finally, proteins and taxonomic assignments are functionally annotated. This can be done through a mix of publicly available databases: UniProt knowledgebase [16], Cluster of Orthologous Groups database [17], the Gene Ontology project [18], and the Kyoto Encyclopedia of Genes and Genomes [19] among others. After annotation this information is all collated with the final objective of elucidating a biochemical description of the target ecosystem.

2 Materials

2.1 Sample Concentration and Size Separation

1. Centramate tangential flow filtration (TFF) system fitted with three 0.1 μm pore size Supor TFF membranes (Pall Corporation).
2. 0.5 M Sodium hydroxide (NaOH).
3. 3 μm pore size polycarbonate isopore membrane (EMD Millipore).

2.2 Sample Lysis

1. Refrigerated centrifuge for 1.5 mL microcentrifuge tubes.
2. 0.5 M triethylammonium bicarbonate buffer (TEAB).

3. Extraction buffer: 0.5 M TEAB, 0.1% sodium dodecyl sulfate (SDS), 1 μL/mL of halt protease inhibitor cocktail.
4. Ultrasonic bath.
5. Water bath.
6. Liquid nitrogen.
7. LoBind microcentrifuge tubes.
8. Disruptor genie bead beater.
9. Glass beads, acid-washed (425–600 μm).

2.3 Acetone Precipitation

1. Benzonase® nuclease.
2. Acetone (−20 °C).

2.4 Protein Quantification

1. Spectrophotometer in the UV range (230 and 260 nm).
2. Quartz cuvette (1 cm light path).

2.5 In-Solution Protein Digest

1. Tris-(2-carboxyethyl)-phosphine (TCEP).
2. Iodoacetamide (IAA).
3. Lyophilized trypsin, MS grade.
4. Hydrochloric acid (HCl).
5. Vacuum centrifuge.

2.6 HPLC

1. Hypercarb porous graphitic column (particle size: 3 μm, length: 50 mm, diameter: 2.1 mm, pore size: 5 μm; Thermo-Dionex).
2. Buffer A: 0.1% (v/v) trifluoroacetic acid (TFA) and 3% (v/v) HPLC grade acetonitrile (ACN) in HPLC-grade water.
3. Buffer B: 0.1% (v/v) TFA and 97% (v/v) ACN in HPLC-grade water.
4. Automated LC vials.
5. Automated LC vial caps.
6. Ultimate 3000 UHPLC (Thermo-Dionex).
7. Chromeleon software (Thermo-Dionex).
8. Fraction collector.

2.7 Mass Spectrometry

1. U3000 RSLCnano LC (Thermo-Dionex).
2. Ultra-high resolution maXis Q-ToF 3G mass spectrometer (Bruker).
3. Data Analysis software (Bruker).
4. Mascot Daemon software (Matrix Science).

3 Methods

3.1 Sample Concentration and Size Separation (See Note 1*)*

1. Environmental samples are concentrated using a Centramate TFF system fitted with three 0.1 μm pore size Supor TFF membranes.
2. After every sample the filter system is sanitized with 0.5 M NaOH and flushed with deionized water.
3. The resulting retentate is discarded.
4. The resulting permeate is filtered with a 3 μm pore size polycarbonate isopore membrane (*see* **Note 2**) in order to obtain fractions dominated by free-living bacteria (<3 μm in size) and algae/particle-associated bacteria (>3 μm in size) (*see* **Note 3**) [3].

3.2 Sample Lysis

1. Each sample is centrifuged at 10,000 × *g* for 15 min at 4 °C and the supernatant is discarded.
2. The resulting cell pellet is washed in 0.5 M TEAB prior to storage, overnight, at −20 °C. If longer storage is needed store at −80 °C.
3. The cell pellet is defrosted, resuspended in 250 μL of extraction buffer, and sonicated in an ice bath for 5 min.
4. The resulting cell suspensions is submitted to five freeze-thaw cycles (each cycle corresponds to 2 min in liquid nitrogen and 5 min in a 37 °C water bath) (*see* **Note 4**) [20].
5. The lysed cell suspension is centrifuged at 15,000 × *g* for 10 min, at 4 °C, and the supernatant is transferred to a LoBind microcentrifuge tube (*see* **Note 5**).
6. The remaining cell pellet is resuspended in 125 μL of extraction buffer and homogenized with glass beads (425–600 μm) for 10 cycles (each cycle corresponds to 2 min homogenization and 2 min on ice).
7. The lysed cell suspension is centrifuged at 15,000 × *g* for 10 min, at 4 °C, and the supernatant is combined with the previous one.

3.3 Acetone Precipitation

1. 1 μL of benzonase® nuclease is added to the sample in order to degrade DNA (*see* **Note 6**).
2. Four volumes of cold acetone are added to the sample and the mixture is stored overnight at −20 °C (*see* **Note 7**).
3. A protein pellet is obtained by centrifugation at 15,000 × *g* for 10 min, at 4 °C, and then vacuum dried (*see* **Note 8**).
4. The protein pellet is resuspended in 100 μL of 0.5 M TEAB and kept at −20 °C until quantification.

3.4 Protein Quantification (See Note 9) [21]

1. Protein samples are thawed and diluted to a range of 6–225 μg/mL (*see* **Notes 10** and **11**).
2. 300 μL (*see* **Note 12**) of 0.5 M TEAB are transferred into a quartz cuvette (*see* **Note 13**) with a 1 cm light path and the buffer is used as a blank.
3. 300 μL of the protein samples, with the appropriate dilution, are transferred into a quartz cuvette with a 1 cm light path.
4. Absorbance is measured at 230 and 260 nm (*see* **Note 14**).
5. Rinse cuvette with deionized H_2O or acetone between measurements.
6. Protein concentrations are calculated using the following equation:
$$\mu g \text{ protein} / \text{mL} = \left(183 \times A_{230} - 75.8 \times A_{260}\right) \times \text{dilution factor}$$

where A_{230} and A_{260} are the absorbances at 230 nm and 260 nm.

3.5 In-Solution Protein Digest

1. Based on the previous quantification an aliquot of the protein sample, equivalent to 200 μg of protein, is transferred into a new LoBind microcentrifuge tube.
2. The protein sample is reduced by adding 20 mM TCEP, at 60 °C, for 30 min.
3. This is followed by alkylation with 10 mM IAA, for 30 min, at room temperature in the dark.
4. Quench unreacted IAA with 20 mM TCEP.
5. The protein sample is digested overnight, at 37 °C, using trypsin 1:40 (trypsin-to-protein ratio) resuspended in 1 mM HCl.
6. The resulting peptide sample is vacuum dried and stored at −20 °C.

3.6 HPLC

1. The first dimensional chromatographic separation, off-line, is performed using a Hypercarb porous graphitic column (particle size: 3 μm, length: 50 mm, diameter: 2.1 mm, pore size: 5 μm) on an Ultimate 3000 UHPLC. Elution is monitored at a wavelength of 214 nm with Chromeleon software.
2. Peptides are resuspended in 200 μL of Buffer A.
3. The peptide mixture is transferred into a vial compatible with the automated LC system with a final volume of 10 μL.
4. One injection of 5 μL is made.
5. A linear gradient of Buffer B ranging from 5% to 60% is used for 110 min within a 120 min running time at a flow rate of 0.2 mL/min.
6. Peptide fractions are collected every 2 min, between 10 and 120 min, using a fraction collector.
7. Peptide fractions are vacuum dried and stored at −20 °C.

3.7 Mass Spectrometry

1. Mass spectrometric analysis is performed on a nano-LC-CSI-MS/MS system with a U3000 RSLCnano LC and a UHR maXis Q-ToF 3G mass spectrometer.
2. Peptide fractions are resuspended in Buffer A and transferred into a vial compatible with the automated LC system with a final volume of 12 μL.
3. Two injections of 5 μL each are made.
4. A linear gradient of buffer B ranging from 4% to 40% is used for 90 min at a flow rate of 0.3 μL/min.
5. On the mass spectrometer, the following settings are specified: endplate Offset −500 V, capillary voltage 1000 V, nebulizer gas 0.4 bar, dry gas 6.0 L/min, and dry temperature 150 °C. Mass range: 50–2200 m/z, at 4 Hz. Data are acquired for positive ions in a dependent acquisition mode with the three most intense double, triple, or quadruple charges species selected for further analysis by tandem mass spectrometry (MS/MS) under collision induced dissociation (CID) conditions where nitrogen is used as collision gas.

3.8 Protein Identification and Quantification

1. MS and MS/MS raw spectra are processed using Data Analysis software and the spectra from each Bruker analysis file (.BAF) were output as a mascot generic file (.MGF).
2. A database for protein identification is created collating all Uniprot entries from organisms with an abundance of >1% in the 16S (prokaryotic) and 18S (eukaryotic) rDNA survey of the sample [14].
3. The subsequent database search is done with Mascot Daemon.
4. The following search parameters are applied: up to one missed cleavage with trypsin, fixed modification of cysteine residues by carbamidomethylation, variable modification of methionine by oxidation, instrument specification ESI Q-ToF, peptide charge: 2+, 3+, and 4+, precursor mass tolerance of ±0.2 Da, and fragment-ion mass tolerance of ±0.02 Da.
5. Only matches above a 95% confidence homology threshold, with significant scores defined by Mascot probability analysis, and a 1% FDR cutoff are considered confidently matched peptides. "Show sub-sets" and "require bold red" are applied on initial Mascot results to eliminate redundancy. The highest score for a given peptide mass is used to identify proteins. Proteins are considered identified only when two or more unique peptides, per protein, are matched.
6. In case a 1% FDR cannot be achieved a two-step search approach can be undertaken [15]. The initial database search is done without any FDR limitation and then followed by a second search with a 1% FDR threshold against a refined database

created by extracting the protein identifications derived from the first search.

7. In Mascot, protein abundance is estimated relatively through the exponentially modified protein abundance index, emPAI [22].

3.9 Functional Classification of Proteins

1. A list of UniProt protein accession numbers are collated from each sample and queried utilizing the UniProt Retrieve/ID mapping tool.
2. Column options "Keywords" and "Gene ontology (biological process)" are selected.
3. Incomplete or ambiguous annotations are manually completed by searching for the individual UniProt protein accession numbers on Pfam and EggNOG.

4 Notes

1. This section is only relevant for metaproteomic studies where samples are dilute (e.g., oligotrophic freshwater environments).
2. When choosing the filter material care should be taken to guarantee compatibility with proteomic workflows.
3. The usage of a larger range of filter sizes will allow for a more comprehensive size separation of the sample.
4. For a quicker thawing procedure water bath temperature can, safely, be increased up to 60 °C.
5. Eppendorf LoBind Tubes maximize sample recovery by significantly reducing sample binding to the surface.
6. The successful application of the nuclease can be confirmed by a substantial decrease in sample viscosity.
7. Acetone precipitation step can be done in as little as 4 h.
8. Acetone should not be allowed to dry completely, otherwise the pellet may be difficult to solubilize.
9. This method utilizes a quartz cuvette and a spectrophotometer. Alternatively, quantification can be done utilizing a NanoDrop (Thermo Scientific).
10. The assay is linear with protein concentration over a range of 6–225 μg/mL. If uncertainty exists regarding the protein concentration range a serial dilution (e.g., 2×, 5×, 10×) is recommended.
11. The assay is insensitive to pH in the range from 6 to 9.
12. In order to minimize sample usage the cuvette should be filled only to the point where the light beam crosses. Volume may

vary according to spectrophotometer model and cuvette design.

13. Cuvette used in the assay must be free from UV-absorbing materials.
14. The same sample aliquot should be used for both measurements.

Acknowledgments

The authors would like to thank the Natural Environment Research Council (NE/J024767/1) and the Technology Strategy Board for funding this work.

References

1. Leary DH, Hervey WJ 4th, Li RW et al (2012) Method development for metaproteomic analyses of marine biofilms. Anal Chem 84(9):4006–4013. https://doi.org/10.1021/ac203315n
2. Leary DH, Hervey WJ 4th, Deschamps JR et al (2013) Which metaproteome? The impact of protein extraction bias on metaproteomic analyses. Mol Cell Probes 27(5–6):193–199. https://doi.org/10.1016/j.mcp.2013.06.003
3. Teeling H, Fuchs BM, Becher D et al (2012) Substrate-controlled succession of marine bacterioplankton populations induced by a phytoplankton bloom. Science 336(6081):608–611. https://doi.org/10.1126/science.1218344
4. Schneider T, Riedel K (2010) Environmental proteomics: analysis of structure and function of microbial communities. Proteomics 10(4):785–798. https://doi.org/10.1002/pmic.200900450
5. Motoyama A, Yates JR (2008) Multidimensional LC separations in shotgun proteomics. Anal Chem 80(19):7187–7193. https://doi.org/10.1021/ac8013669
6. Bereszczak JZ, Brancia FL (2009) Offline and online liquid chromatography mass spectrometry in quantitative proteomics. Comb Chem High Throughput Screen 12(2):185–193. https://doi.org/10.2174/138620709787315418
7. Fenn JB, Mann M, Meng CK et al (1989) Electrospray ionization for mass spectrometry of large biomolecules. Science 246(4926):64–71. https://doi.org/10.1126/science.2675315
8. Hillenkamp F, Karas M, Beavis RC et al (1991) Matrix-assisted laser desorption/ionization mass spectrometry of biopolymers. Anal Chem 63(24):1193A–1203A. https://doi.org/10.1021/ac00024a716
9. Yates JR, Ruse CI, Nakorchevsky A (2009) Proteomics by mass spectrometry: approaches, advances, and applications. Annu Rev Biomed Eng 11:49–79. https://doi.org/10.1146/annurev-bioeng-061008-124934
10. Muth T, Benndorf D, Reichl U et al (2013) Searching for a needle in a stack of needles: challenges in metaproteomics data analysis. Mol BioSyst 9(4):578–585. https://doi.org/10.1039/c2mb25415h
11. Perkins DN, Pappin DJ, Creasy DM et al (1999) Probability-based protein identification by searching sequence databases using mass spectrometry data. Electrophoresis 20(18):3551–3567. https://doi.org/10.1002/(SICI)1522-2683(19991201)20:18<3551::AID-ELPS3551>3.0.CO;2-2
12. Craig R, Beavis RC (2004) TANDEM: matching proteins with tandem mass spectra. Bioinformatics 20(9):1466–1467. https://doi.org/10.1093/bioinformatics/bth092
13. Pible O, Armengaud J (2015) Improving the quality of genome, protein sequence, and taxonomy databases: a prerequisite for microbiome meta-omics 2.0. Proteomics 15(20):3418–3423. https://doi.org/10.1002/pmic.201500104
14. Russo DA, Couto N, Beckerman AP et al (2016) A metaproteomic analysis of the response of a freshwater microbial community

under nutrient enrichment. Front Microbiol 7:1172. https://doi.org/10.3389/fmicb.2016.01172

15. Jagtap P, Goslinga J, Kooren JA et al (2013) A two-step database search method improves sensitivity in peptide sequence matches for metaproteomics and proteogenomics studies. Proteomics 13(8):1352–1357. https://doi.org/10.1002/pmic.201200352
16. Apweiler R, Bairoch A, Wu CH et al (2004) UniProt: the Universal Protein knowledgebase. Nucleic Acids Res 32(database issue): D115–D119. https://doi.org/10.1093/nar/gkh131
17. Tatusov R, Fedorova N, Jackson J et al (2003) The COG database: an updated version includes eukaryotes. BMC Bioinformatics 4(1):1–14. https://doi.org/10.1186/1471-2105-4-41
18. Ashburner M, Ball CA, Blake JA et al (2000) Gene ontology: tool for the unification of biology. Nat Genet 25(1):25–29. https://doi.org/10.1038/75556
19. Kanehisa M, Goto S (2000) KEGG: Kyoto Encyclopedia of Genes and Genomes. Nucleic Acids Res 28(1):27–30. https://doi.org/10.1093/nar/28.1.27
20. Ogunseitan OA (1993) Direct extraction of proteins from environmental samples. J Microbiol Meth 17(4):273–281. https://doi.org/10.1016/0167-7012(93)90056-N
21. Kalb VF Jr, Bernlohr RW (1977) A new spectrophotometric assay for protein in cell extracts. Anal Biochem 82:362–371. https://doi.org/10.1016/0003-2697(77)90173-7
22. Ishihama Y, Oda Y, Tabata T et al (2005) Exponentially modified protein abundance index (emPAI) for estimation of absolute protein amount in proteomics by the number of sequenced peptides per protein. Mol Cell Proteomics 4(9):1265–1272. https://doi.org/10.1074/mcp.M500061-MCP200

Part II

Dealing with Proteomics Data in a Big Data Era

Chapter 11

Peptide-to-Protein Summarization: An Important Step for Accurate Quantification in Label-Based Proteomics

Martina Fischer, Thilo Muth, and Bernhard Y. Renard

Abstract

Quantitative MS/MS-based measurements are assessed at the peptide spectrum level and substantial variance is frequently observed for any given protein. Protein quantification requires a peptide-to-protein summarization step. This important step has been little investigated and most strategies only rely on quantitative spectrum values, ignoring a wealth of additional feature information is available for peptide spectra.

In this chapter, we discuss summarization methods that can be applied for label-based protein quantification. In particular, we focus on strategies using peptide spectrum characteristics in addition to quantitative values for protein abundance inference. We highlight significant relations of spectrum features and quantification accuracy to assess the reliability of spectra and the development of a correction. As a result, spectra of lower quality are identified, their impact is minimized and overall protein quantification is improved. Here, we investigate different peptide features in detail, emphasize the benefits of integrating spectrum feature information, and provide recommendations on the usage of the methods.

Key words Protein quantification, Peptide-to-protein summarization, Isobaric mass tagging, iTRAQ, TMT, Label-based proteomics, Quantitative proteomics, Peptide features

1 Introduction

Mass spectrometry-based proteomics methods are routinely applied for the analysis of proteins in any biological sample under varying conditions and time points. The nature and identity of proteins expressed at a given time convey information about proteome dynamics, function, and homeostasis, including the synthesis and degradation of proteins, or the formation of protein complexes. It is also informative to retrieve quantitative data about relative protein abundances between different samples. Instead of time-consuming individual protein testing, modern quantitative proteomics workflows enable researchers to calculate the abundance of thousands of proteins at the same time. While peptide or protein identification is relatively straightforward using search

Caroline A. Evans et al. (eds.), *Mass Spectrometry of Proteins: Methods and Protocols*, Methods in Molecular Biology, vol. 1977, https://doi.org/10.1007/978-1-4939-9232-4_11,

algorithms matching experimental tandem mass spectra against sequence databases, quantitative proteomics workflows are however commonly more complex and require a complete range of highly diverse methodologies which are applied at the experimental as well as the data processing level.

1.1 State of the Art

In general, protein quantification can be classified into the two main categories of label-free and label-based proteomic methods. Label-free quantitative methods are based on spectral counting or ion intensities and gain increasing popularity because of cost-efficient and non-limited sample processing [1], but they require a high reproducibility of the experimental setup. Label-based workflows are still widely used in the field which is mainly due to their accuracy and speed given by the capability of performing multiplexed analyses (see below).

Label-based quantification can be subdivided into in vivo (metabolic) and in vitro (chemical and enzymatic) labeling strategies. A common and robust in vivo approach presents stable isotope labeling of amino acids in cell culture (SILAC) [2]. SILAC relies on metabolic labeling, in which heavy isotope-labeled (^{13}C) amino acids, such as lysine and arginine, are incorporated into proteins of cells within a growth medium. For chemical in vitro labeling, isobaric mass tags for absolute and relative quantification (iTRAQ) [3] and tandem mass tags (TMT) [4] are common techniques used for the quantification of proteins. In these techniques, tryptic peptides are labeled with chemical (mainly amine-reactive) tags which produce reporter ions of different mass to be detected at the MS/MS level, namely in course of peptide fragmentation. The main idea is to prepare each sample with a different type of isobaric mass tag for subsequently performing a multiplexed analysis of various samples simultaneously within a single experimental run (e.g., 6-plex for TMT and 8-plex iTRAQ labeling, respectively). While the fragment ions at a higher mass-to-charge ratio are used for the peptide identification, the signals within the reporter tag region then provide quantitative information in correlation to the relative abundance of the peptides in the samples. The relative expression level for each protein is elucidated by integrating the quantitative information of the assigned peptides.

Several factors affect the protein quantification when using isobaric labeling techniques and problems in accuracy have been reported in various studies [5–8]. At the technical side, the most severe issues arise from limited labeling efficiency, isotope impurity, dynamic range of reporter ion intensity, co-elution of peptides, and artifactual peaks within the reporter region [9–11]. Among them, precursor co-isolation presents a prominent mechanism through which quantification accuracy deteriorates: this effect has been frequently reported of leading to an underestimation in the ratio of protein abundance differences in analyzed samples [6, 12, 13]. In this context, dedicated computational methods have been established to address the aforementioned issues, for instance, by performing intensity calculation, impurity correction, accuracy, and error estimation on given isotope data [14–16]. In addition, complete software packages have been proposed by which quantitative isobaric tag data can be processed and interpreted in detail [17–19]. To control the variation arising when considering several samples, (technical, experimental, and biological) replicates and spike-in proteins should be included in the experimental design [5, 10].

For each individual sample, an essential step is to derive quantitative information at the protein level from the assigned peptide spectra. A summarization strategy is necessary to estimate the protein ratio by inferring the intensities of the corresponding peptide signals. Ideally, all assigned peptide-spectrum matches of a single protein would report similar intensity values. In reality, however, accuracy of protein quantification is strongly affected by the heterogeneity at the peptide level because of random and systematic bias [6]. As illustrated in Fig. 1a, enormous variance heterogeneity can be observed for quantitative information of unique peptides assigned to a single protein. To address this issue at the level of data analysis, several methods have been developed in the past which account for reported intensity-dependent effects, such as low-intensity peptide filtering [20], peptide intensity weighting [21], or variance stabilization [6]. Moreover, different statistical methods have been used for estimating the protein ratio from multiple peptides in various quantification software tools, including mean, median, or weighted mean calculation [15, 21, 22]. Further methods develop tailored error and noise models according to the underlying protein abundance ratio distributions [23, 24]. In general, these summarization methods rely on the absolute intensity as the only relevant feature within the MS/MS spectrum, although it was reported in the literature that other features severely affect individually measured peptides within a protein, such as charge state [25], posttranslational modifications, or splice variants [26].

In the past, several ideas have been proposed to address this. For example, Lai et al. [27] suggested using filters for removing potentially unreliable peptides based on peptide frequency, retention time, or lack of reproducibility. Tang et al. [28] focus on improving quantification by incorporating peptide detectability, whereas IDEAL-Q [29] includes retention time detection. Diffacto [30] relies on a factor analysis model for summarizing peptide levels results and thereby downweight unreliable measurements, while Goeminne et al. [31] introduce a robust ridge regression approach. All combined, these examples show that various features

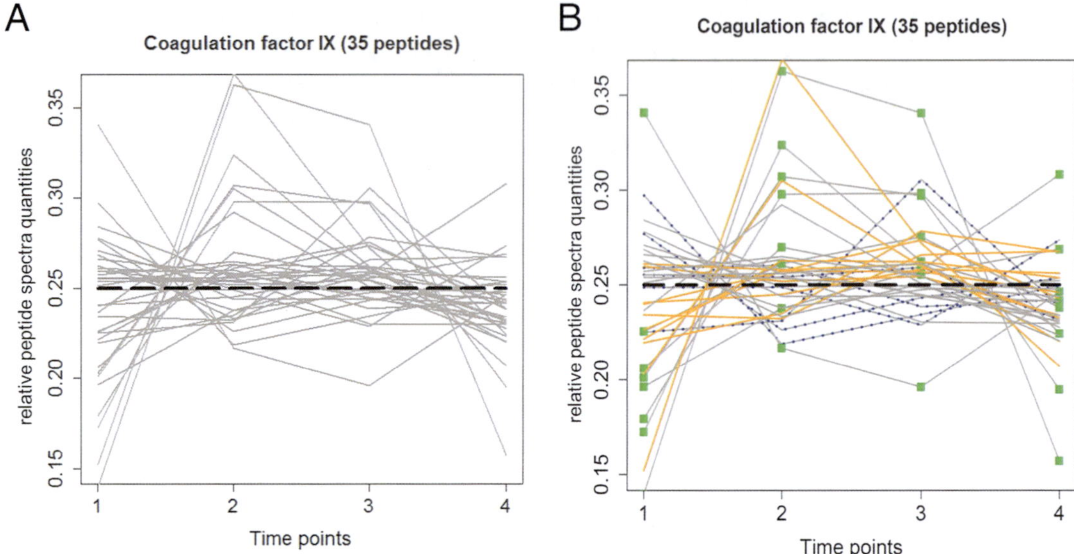

Fig. 1 Variation and heterogeneity among spectrum quantification measurements: example of the human coagulation factor IX (accession P00740) and its 35 MS2 spectra. Every line corresponds to relative quantities of one peptide spectra. The underlying true quantities are marked by a dashed black line. (**a**) Raw data and (**b**) data annotated with peptide features: a redundant spectrum group is highlighted in orange, spectra with charge state greater than two are marked by green boxes, and modified spectra are shown by dotted lines. (Reproduced from [27], with permission from Oxford University Press)

and integration schemes have clear merits. Further investigation may help to identify ways to integrate large collection of features that represent different aspects of data acquisition into a single method. That again may then serve to make protein summarization and quantification more robust in both label-based and label-free proteomics even for small numbers of peptide spectra matches.

These aspects may be of particular relevance when regarding comparative studies contrasting the quantification of proteins across experimental conditions. The estimation of protein abundance always has an immediate impact on subsequent differential assessment. Errors in quantification and underestimation of technical variation may result in the incorrect detection of differentially abundant proteins. For accurate differential assessment, it is crucial to reduce the variance of given peptide quantities, this can be achieved by weighting spectra according to feature information. While numerous tools have been introduced for comparative omics [32–34], it has also been shown that they need to be closely tailored to the underlying data and also quantification algorithms [35].

In this contribution, we describe a method that takes into account the influence of particular features of each peptide-spectrum match, including charge state, sequence length, identification score, precursor mass, and distance between single and multiple recorded spectra per peptide, on protein quantification. We discuss our previously published algorithm iPQF (isobaric Protein Quantification based on Features), which estimates protein quantities based on the listed peptide-spectrum features and quantitative values [36]. We will use it as an example to highlight the relation of spectra features and quantification accuracy. A technical description and a detailed example workflow are provided to enable comprehensible usage of the method and a section with guidelines and recommendations are given.

1.2 Peptide Features

Quantitative MS/MS-based measurements are obtained at the peptide spectrum level. Thereby, substantial variance is commonly observed for the set of spectra assigned to the same protein, as illustrated in Fig. 1a. How can we explain this heterogeneity? Do spectra which diverge from the ground truth exhibit certain feature characteristics? A wealth of information is available for peptide spectra beyond the observed quantitative measures.

Integrating more information into our example in Fig. 1b reveals that many of the spectra which diverge significantly from the ground truth hold higher charge states or show specific modifications. Further, we can observe that among a large group of spectra, all matched to the same peptide sequence, the majority follows one common quantification trend and reflects the course of the ground truth, while few of the spectra deviate from the group as well as from the ground truth.

The objective here is to examine how peptide features contribute to the variance heterogeneity. Therefore, we systematically investigate individual features and their impact on the accuracy of quantification. The seven peptide features considered are (1) charge state, (2) sequence length, (3) peptide mass, (4) identification score, (5) modification state, (6) absolute ion intensities, and (7) heterogeneity within redundantly measured spectra derived from the same precursor. We explain and visualize all these features exemplarily on a published dataset by Hultin-Rosenburg et al. [8], corresponding to a human cell lysate labeled with an 8-plex iTRAQ and analyzed on a LTQ Orbitrap Velos.

Depending on its sequence length and chemical properties, each protein within a biological sample is enzymatically digested (e.g., using tryptic cleavage) into a defined set of peptides. Several MS/MS scans are commonly obtained for the same peptide in an experiment; hereafter, we refer to those as "redundant" spectra. Hence, a protein is generally defined by different groups of redundant spectra and also individual "unique" spectra (i.e., corresponding to peptides with exactly one MS/MS event). In fact, datasets are predominantly characterized by large numbers of redundant spectra. As redundant spectra can be interpreted as repeated measurements of a peptide fragment underlying the same conditions in the MS experiment, high similarities in quantitative measurements are expected. Hence, valuable information concerning reliability of measurements can be retrieved from these spectrum groups. We introduce a distance metric which computes the mean Euclidian distance of each peptide spectrum quantity to the measurements of all other spectra belonging to the same group. Besides the redundant spectra groups, we also define one group in which all uniquely measured peptides are pooled. The general assumption is that the true quantification values better approximated by a majority of spectra with highly similar measurements. A peptide spectrum which exhibits quantities diverging from all other measurements given for spectra within its group is deemed to be less reliable. This way, individual spectra or even an entire group of spectra with overall high distance values can be marked and ranked compared to other spectrum groups.

We tested the impact of heterogeneity in spectrum groups and the proposed metric on different MS2 and MS3 datasets [8, 23, 24, 37] with available ground truth. Altogether, these studies confirmed a strong positive correlation of quantitative spectra distance to actual quantification error. Figure 2a and b shows how the quantification error increases with a spectrum quantitatively diverging from its assigned group. As the majority of spectra are expected to spread around the true underlying quantification values, highest density is found for low values of error and distance within each group. Most importantly to note is that almost no spectrum holds a small euclidean distance with a large quantification error.

Further, the absolute peak intensity of spectra plays a crucial role and has been the most investigated feature so far. Several studies demonstrate that low signal intensity spectra are more prone to noise and error and significantly contribute to variance heterogeneity [6, 38, 39]. One possibility is that ion counts are measured as whole numbers by MS instruments [40, 41]. But quantification accuracy is shown to improve for spectra with high absolute intensity values. In Fig. 2c, the mean absolute intensity of a spectrum across all labels is correlated with the spectrum quantification error, a bend clearly indicates a threshold from which spectra quantities become more reliable.

Moreover, spectra reliability is also fundamentally determined by the previous step of peptide identification. Quantification will be significantly biased in case a spectrum was not assigned to the correct protein. Different database-driven identification algorithms provide scores indicating the goodness of a match between a spectrum and a reference peptide sequence. However, depending on the scoring system, its computation, and the underlying sequence database, the relation between identification score and quantification error can be complex and may vary across systems. In our example dataset, the identification score refers to the posterior error probability (PEP) and Fig. 2d displays a dominant value

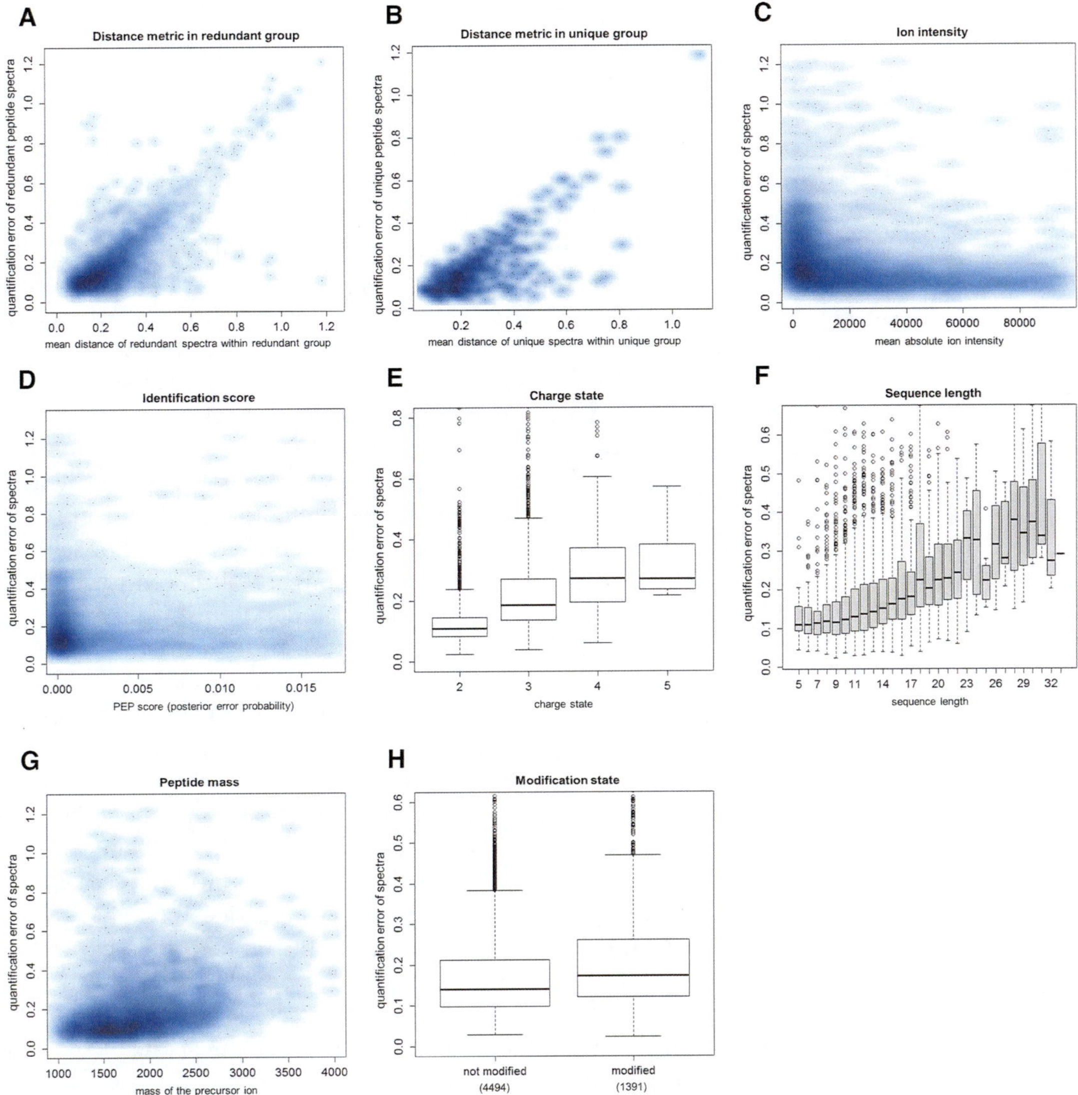

Fig. 2 Correlations between different peptide features and the quantification error of peptide spectra. The quantification error computed for each spectrum in the dataset [8] refers to an euclidean distance of the measured relative spectra quantities to the expected quantities according to the given ground truth. Significant correlations are observed for the presented features indicating an impact on quantification accuracy. (Reproduced from [27], with permission from Oxford University Press)

concentration of small PEPs with small quantification errors. But here no overall correlation trend is visible and the used score does not serve as a strong supportive feature in this case.

Although the score itself may not necessarily provide good indication on spectra reliability, there are several other features which directly and indirectly impact identification such as the charge state, the mass, and the sequence length of a peptide. Peptides with higher charge states can result in a greater number of possible fragments holding different amounts of charges, consequently requiring a larger peptide search space along with the risk of more false-positive identifications. Likewise, longer peptides yield a greater amount of b- and y-ions in the MS/MS scan accompanied with more potential errors, partly due to incomplete fragmentation, which can complicate spectrum to peptide inference. Figure 2e reveals the decrease of quantification accuracy with increasing charge state, particularly for the common charge states two, three, and four. As expected, peptides with increasing sequence length from 5 to 30 amino acids are shown to correlate with increasing quantification error in Fig. 2f. A positive correlation coefficient of 0.38 also indicates a significant decrease of quantification accuracy with increasing peptide mass, though less visible in Fig. 2g due to a large diversity of values.

An additional feature concerns modifications of peptides which are frequently occurring in enzymatic or sample preparation reactions and correspondingly challenge the peptide search. For several protein cases, shifted quantification values are observed for underlying modified spectra. So far, we did not distinguish the impact of specific types of modifications on quantification accuracy, but only consider the two states of modified and unmodified peptides. Figure 2h confirms a loss of accuracy for the group of modified peptides.

Overall, we here examined different individual features which are related to spectra reliability and which are shown to have a direct or indirect impact on quantification accuracy. Many of the features are intercorrelated in some way; thus, longer sequences exhibit a higher mass and the number of possible charges is correspondingly increasing. Likewise, shorter peptide sequences commonly hold fewer modifications and enable easier spectrum-to-peptide identification, receiving better scores accordingly. Hence, some features have strong interdependencies, while others are less correlated. However, neither of the individual features is redundant nor is one feature significant enough to serve as exclusive estimator for spectrum reliability and for corresponding quantification inference. Consequently, using a combination of features is key to compensate for the shortcomings of any single feature. All in all, the probability that a peptide spectrum is misclassified by a large set of features simultaneously will be minimized.

1.3 iPQF Algorithm

The iPQF algorithm (isobaric Protein Quantification based on Features) represents a peptide-to-protein summarization method [36]. The abundance of a protein is computed by integrating peptide spectrum features with quantitative values. Characteristics of spectra are used to assign weights to peptide spectra, which indicate reliability for protein quantification.

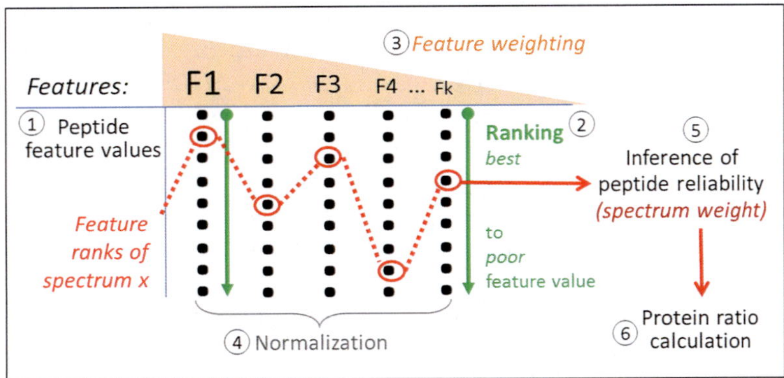

Fig. 3 Schematic overview of the approach taken by iPQF to derive accurate protein quantification: (1) feature assessment, (2) spectrum ranking, (3) feature weighting, (4) normalization, (5) spectrum reliability, (6) protein quantification. (Reproduced from [27], with permission from Oxford University Press)

The abundance of each protein is calculated according to the following six steps (Fig. 3), which will be further illustrated in Subheading 3.2:

1. Feature assessment.
2. Spectrum ranking per feature.
3. Feature weighting.
4. Normalization.
5. Reliability analysis of each spectrum.
6. Protein quantification.

2 Materials

The iPQF algorithm is implemented in the software environment R and is part of the existing R/Bioconductor package *MSnbase* [42].

1. GNU R (version ≥ 3.1.3, https://www.r-project.org/).
2. The MSnbase package can be installed by running the following commands:

   source("https://www.bioconductor.org/biocLite.R")

   biocLite("MSnbase").

MSnbase is a comprehensive package offering a range of data processing, visualization, and quantification functions for MS-based proteomics data. Detailed information on these functions is available in the vignette and the manual on the corresponding Bioconductor package website (https://bioconductor.org/packages/release/bioc/html/MSnbase.html). The user can start from raw MS/MS spectrum data by importing *XML*-based (mzXML, mzML) [43, 44] or *MGF* formats, or by directly integrating already available MS quantitation data in spreadsheet format.

The package provides all necessary spectrum preprocessing steps such as quality control, filtering of low-intensity peaks, trimming, impurity correction, and missing value imputation. The user can choose from three different methods to obtain quantitation on fixed peaks in the spectra: (1) area under the peak, (2) height at the apex of the peak, or (3) summation of all intensities corresponding to the peak. Subsequent normalization of samples can be conducted. As a result, quantities of individual peptide spectra are obtained.

3 Methods

3.1 Feature Weights

Based on the general correlations described in the Introduction between features and quantification error, weights are attributed to each feature. They are computed, based on a robustness analysis, as the squared order of the features redundant (*see* **Note 1**):

1. Unique-distance metric (weight 49).
2. Charge state (weight 36).
3. Ion intensity (weight 25).
4. Sequence length (weight 16).
5. Identification score (weight 9).
6. Modification state (weight 4).
7. Precursor mass (weight 1).

This means, for instance, that the unique-distance metric is single most important parameter allowing the user to take a guess at what spectrum is reliable or not.

3.2 Summarization Steps

The iPQF algorithm considers each protein in a dataset individually and evaluates the abundance of each protein based on the information of the peptide spectra assigned said protein. The quantification of one protein is independent of the quantification of other proteins in the dataset. Here, we want to demonstrate all algorithmic steps by taking a specific protein as an example. For this purpose, we select a single protein from the dataset of Hultin-Rosenberg et al. [8], the FAM49B protein, which is involved in platelet degranulation (IPI:IPI00303318.2). It was identified by eight spectra in the MS/MS experiment as shown in Fig. 4a, corresponding to two separate peptides.

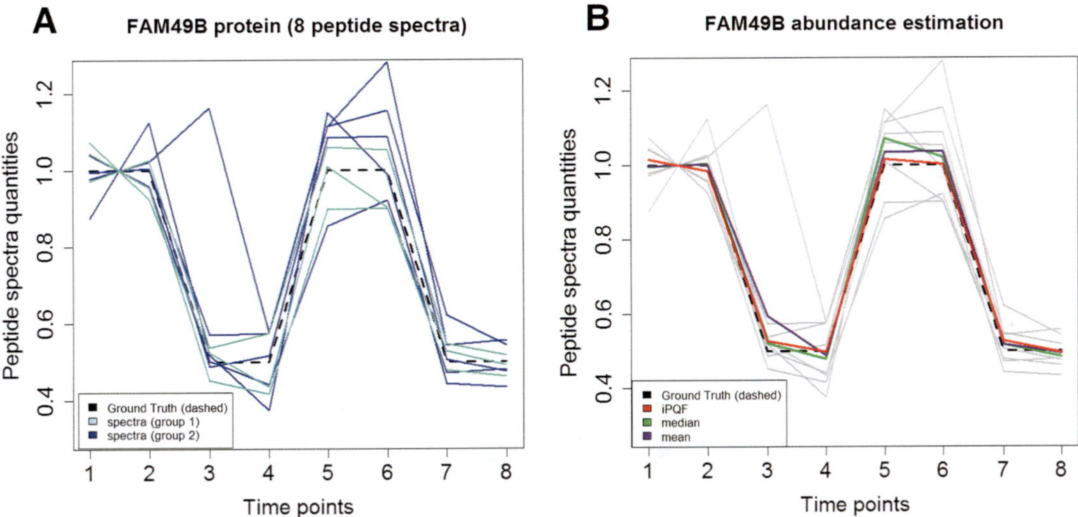

Fig. 4 Peptide spectra abundance profile of protein FAM49B. (**a**) Relative quantities of the eight assigned peptide spectra for the eight time points of an 8-plex iTRAQ are displayed and a significant heterogeneity is observed. (**b**) Three different summarization methods for protein abundance inference are shown

1. Feature assessment. For all peptide spectra assigned to the considered protein, values of the seven presented features (charge state, sequence length, peptide mass, identification score, modification state, absolute ion intensities, and distance metric on heterogeneity) are given. The values are displayed in a matrix with peptide spectra referring to rows and features to columns (Table 1).

   Let's consid.er the eight spectra corresponding to two different peptide sequences and each peptide was measured by several MS/MS scans (*see* **Note 2**). Hence, two groups of redundant spectra are recorded, holding three and five spectra, respectively; the proposed distance metric is calculated for each spectrum within its group. We observe that quantitative measures of spectra representing peptide 1 reveal to be more closely aligned, while the quantities given for spectra of peptide 2 prove to be less homogenous, with spectrum ID 3961 diverging particularly far within its group.

   Information regarding the six other features is retrieved from the peptide identification output file. Overall, two different charge states are present, notably a mix of charges is observed within the groups of redundantly measured spectra. The absolute ion intensity values spans two orders of magnitude. The diversity in sequence length is necessarily less remarkable (15 and 19 amino acids). Correspondingly, the mass of the precursor ion, computed by multiplication of the m/z value and the

Table 1
Feature assessment

Spectrum		Features						
ID	Spectrum group	Distance metric (unique, redundant spectra	Charge state	Ion intensity	Sequence length	Identification score	Modification	Mass
908	1	0.2111538	3	17339.48	15	4.43×10^{-4}	0	1929.07
909	1	0.2871840	2	90938.596	15	8.56×10^{-4}	0	1928.06
910	1	0.2287298	3	27548.177	15	1.80×10^{-3}	0	1929.07
3957	2	0.4889893	2	32940.998	19	7.13×10^{-8}	1	2487.21
3958	2	0.4181052	2	8339.638	19	1.44×10^{-6}	1	2503.20
3959	2	0.5168905	3	4650.797	19	2.82×10^{-6}	1	2488.21
3960	2	0.4181052	3	10022.587	19	3.65×10^{-6}	1	2504.21
3961	2	0.7404597	3	10213.959	19	7.79×10^{-6}	1	2488.21

Table 2
Spectrum ranking

Spectrum		Features						
ID	Spectrum group	Distance metric (unique, redundant spectra	Charge state	Ion intensity	Sequence length	Identification score	Modification	Mass
908	1	1	6	4	2	3	2	3
909	1	3	2	1	2	2	2	1
910	1	2	6	3	2	1	2	2
3957	2	6	2	2	6	8	6	4
3958	2	4.5	2	7	6	7	6	7
3959	2	7	6	8	6	6	6	5
3960	2	4.5	6	6	6	5	6	8
3961	2	8	6	5	6	4	6	6

charge state, reflects the correlation to the sequence length. Further, the identification score, in this example referring to the PEP scores (posterior error probability) recorded by Mascot 2.2, indicates diverse identification reliability. Modifications are exclusively found for the second spectrum group, additionally differentiating the spectra.

2. Spectrum ranking per feature (Table 2).

   For each feature, the values are ranked from most to least reliable based on knowledge acquired in the feature-error-correlation studies (*see* Fig. 2), with a rank of 1 being the highest and most reliable. If two spectra hold equal feature values, they receive the same rank. Thus, in this example, the smallest distance metric and best rank is found for spectrum ID 908, while the identification score supports quantities given for the third spectrum holding the best score, and measures of spectrum ID 909 are highlighted by the feature of ion intensity. As only two different charge states exist, ranks are equally split among them. Hence, overall, each peptide spectrum may receive a different rank and quantification reliability by each feature. However, the assumption is that spectra with accurate quantities close to the true quantification values will primarily receive high ranks and support by a majority of features.

3. Feature weighting (Table 3)

   The features are weighted according to their expected individual impact on quantification accuracy. A feature order and robust default weighting is given which is derived from previous studies (*see* Subheading 3.1). Based on our prior analysis, the distance metric is the single most significant feature to distinguish spectra, followed by the charge state, etc., thus enabling more general differentiation and the absolute ion intensity improving the fine ranking of spectra. Here is how it's done: the ranks in each feature column are multiplied by the corresponding feature weights defined in Subheading 3.1 (*see* **Note 3**).

4. Normalization (Table 4)

   In this step, all values are divided by the total number of peptide spectra assigned to the considered protein.

5. Reliability analysis (Table 5)

   Each peptide spectrum holds several feature ranks and each rank reflects the quality of the spectrum for quantification. In this step, all ranks of a spectrum are combined to calculate its overall reliability. Therefore, the ranks of each spectrum are summed up and divided by the sum of the feature weights, resulting in an average rank per spectrum. A low average rank in Table 5 corresponds to a higher quantification reliability of the spectrum.

   Average ranks are converted into a peptide spectrum weight, with higher values assigned to highly reliable spectra; weights are normalized to sum to one. As a result, peptide spectra having weights close to one form the group of highly reliable spectra which are predominantly used for inference of the protein abundance, while spectra with weights decreasing to zero receive less confidence and contribute with smaller part to the protein quantification. In this example, the quantitative measurements given for spectra of peptide 1 gained the highest weights and will primarily determine the protein abundance estimation (Table 6).

6. Protein quantification. For each protein, the relative reporter ion intensities of the assigned spectra serve as input (Table 7). A classic weighted mean approach is computed according to the given feature-based weights of the peptide spectra (*see* **Note 4**) and this leads to the final quantifications: 1.02 (label 113), 0.98 (label 114), 0.53 (label 115), 0.5 (label 116), 1.01 (label 117), 1.00 (label 118), 0.53 (label 119), and 0.49 (label 121). Figure 4b shows that an accurate estimation of the underlying relative abundance of the protein is obtained (*see* **Note 5**).

   There are some limitations iPQF cannot deal with. The first limitation lies with the interference effect caused by co-eluting peptides which results in quantitative underestimation and ratio compression toward one. Only an additional MS3 frag-

Table 3
Feature weighting

Spectrum		Features						
		Feature order (weighting by squared order value)						
		7	6	5	4	3	2	1
ID	Spectrum group	Distance metric (unique, redundant spectra	Charge state	Ion intensity	Sequence length	Identification score	Modification	Mass
908	1	49	216	100	32	27	8	3
909	1	147	72	25	32	18	8	1
910	1	98	216	75	32	9	8	2
3957	2	294	72	50	96	72	24	4
3958	2	220.5	72	175	96	63	24	7
3959	2	343	216	200	96	54	24	5
3960	2	220.5	216	150	96	45	24	8
3961	2	392	216	125	96	36	24	6

Table 4
Normalization

Spectrum		Features						
ID	Spectrum group	Distance metric (unique, redundant spectra	Charge state	Ion intensity	Sequence length	Identification score	Modification	Mass
908	1	6.125	27	12.5	4	3.375	1	0.375
909	1	18.375	9	3.125	4	2.25	1	0.125
910	1	12.25	27	9.375	4	1.125	1	0.25
3957	2	36.75	9	6.25	12	9	3	0.5
3958	2	27.5625	9	21.875	12	7.875	3	0.875
3959	2	42.875	27	25	12	6.75	3	0.625
3960	2	27.5625	27	18.75	12	5.625	3	1
3961	2	49	27	15.625	12	4.5	3	0.75

Table 5
Overall reliability of peptide spectra

Spectrum ID	908	909	910	3957	3958	3959	3960	3961
Sum of ranks	54.375	37.875	55	76.5	82.1875	117.25	94.9375	111.875
Average rank	0.3883929	0.2705357	0.3928571	0.5464286	0.5870536	0.8375	0.678125	0.7991071

Table 6
Weights for each spectrum

Spectrum ID	908	909	910	3957	3958	3959	3960	3961
Weight	0.20536867	0.29214413	0.20238153	0.11294850	0.09362170	0.01449759	0.05688052	0.02215736

Table 7
MS/MS label-based quantifications

Spectrum ID	Label 113	Label 114	Label 115	Label 116	Label 117	Label 118	Label 119	Label 121
908	1.07413	0.925867	0.524194	0.434084	1.00971	0.902939	0.478253	0.461447
909	0.973043	1.02696	0.537628	0.574595	1.05896	1.05198	0.546601	0.516376
910	1.04526	0.95474	0.451316	0.415578	0.897413	0.901616	0.528934	0.490008
3957	0.994696	1.0053	0.486815	0.517545	1.08323	1.08868	0.5055	0.473693
3958	0.996895	1.00311	0.571591	0.575438	1.11389	1.15476	0.621393	0.542488
3959	0.87452	1.12548	0.518196	0.374228	1.14955	0.989283	0.441619	0.432693
3960	1.04146	0.958543	0.501505	0.439843	0.854987	0.922625	0.471108	0.478929
3961	0.97839	1.02161	1.1633	0.57582	1.11344	1.28098	0.542649	0.556815

mentation step can properly circumvent this effect. Another limitation has to do with either missing values (which be more or less frequent depending on the MS equipment) and shared peptide spectra (i.e., peptide sequences can be traced back to several proteins).

3.3 Using iPQF in R

The iPQF method is included as part of the MSnbase package.

1. Load the MSnbase package (*see* Materials for installation instructions):

   library("MSnbase")

2. Load the example dataset "msnset2"

   data(msnset2)

   Using the following commands

   class(msnset2)

   dim(msnset2)

   we learn that the dataset consists of 614 peptide-to-spectrum matches and four quantifications. The dataset has class MSnSet; real data can, for instance, be loaded into R using the readMSnSet() function.

3. Summarization is carried out by the combineFeatures() function:

   ```
   res = combineFeatures(msnset2,
             groupBy=fData(msnset2)$accession,
             redundancy.handler="unique",
             fun="iPQF)
   ```

   The "groupBy" argument specifies what groups of spectra the summarization must consider to compute protein-level quantifications; the accession number of the protein is used. The "redundancy.handler" is set to unique; this means that spectra that are matched to several proteins should be discarded from the calculations (*see* **Notes 6** and 7). The "fun" argument is set to "iPQF" to use the iPQF method described here. Expression levels can be interrogated through

   exprs(res)

   (For differential analysis, *see* **Note 8**.)

4 Notes

1. The combineFeatures() function can pass arguments to the iPQF procedure. Among other arguments, there is the "feature.weight" argument, which allows the user to prioritize features differently should the data display other feature-error-correlations.
2. Methods such as iPQF have been proven to be particularly useful when the number of spectra for a given protein is small (1–6) or medium (7–15). We recommend a minimum of three

peptide spectra per protein to ensure meaningful protein quantification.

3. If the user intends to add new features, an internal correlation check to existing features should be conducted to avoid enhancement of highly correlated features.
4. Peptide spectra quantities contribute to the overall protein abundance estimation according to their weights, a classic weighted mean approach is applied ("pure iPQF"). Further, quantitative estimates derived by the Median Polish approach can be additionally integrated ("combined iPQF").
5. Assuming the spectrum weights are contained in a row vector W and the quantifications are contained in a matrix Q (whose entries, for instance, would be the values of Table 7); the final quantifications are simply given by $W \cdot Q$, where $\cdot$ refers to the matrix multiplication.
6. Further, the method of iPQF can be combined with a pure and more sophisticated quantitative summarization method to join strength of two approaches. This can be prompted by the method.combine argument, here, a combination with the method of Median Polish [42] is provided. Beyond the standard command call, the advanced R user can also extract the computed spectra weights, which refer to spectra reliability for quantification, and employ it in a different setting or purpose.
7. Specific arguments to iPQF are low.support.filter, which enables to filter proteins which are supported by less than three spectra, and ratio.calc, which defines the computation of the peptide ratios. Generally, we recommend using relative spectra intensities in order to estimate protein ratios. Relative intensities are more robust, while the variance in absolute intensities can be large. Ratios can be either calculated by dividing the absolute intensity of one label by the summed intensities of all iTRAQ labels, which refers to a robust approach with greater label independence, or by dividing all spectra intensities by the spectra intensity of a specific iTRAQ label. This choice depends on the question addressed by the experiment, the latter approach may be reasonable, e.g., if one label serves as baseline.
8. Accurate protein quantification in a sample forms the base for successful differential quantification. Correct detection of significant abundance change of proteins between samples and a precise inference of log-fold changes is determined by the accuracy of estimations given in each sample. Thus, a method such as iPQF can indirectly help to reduce the number of false-positives resulting from a differential analysis. However, the choice of a suitable differential test remains strongly dependent on the experimental technique applied, corresponding data characteristics, and number of replicates available.

References

1. Neilson KA, Ali NA, Muralidharan S et al (2011) Less label, more free: approaches in label-free quantitative mass spectrometry. Proteomics 11:535–553
2. Ong SE, Blagoev B, Kratchmarova I et al (2002) Stable isotope labeling by amino acids in cell culture, SILAC, as a simple and accurate approach to expression proteomics. Mol Cell Proteomics 1:376–386
3. Ross PL, Huang YN, Marchese JN et al (2004) Multiplexed protein quantitation in *Saccharomyces cerevisiae* using amine-reactive isobaric tagging reagents. Mol Cell Proteomics 3:1154–1169
4. Thompson A, Schafer J, Kuhn K et al (2003) Tandem mass tags: a novel quantification strategy for comparative analysis of complex protein mixtures by MS/MS. Anal Chem 75:1895–1904
5. Gan CS, Chong PK, Pham TK et al (2007) Technical, experimental, and biological variations in isobaric tags for relative and absolute quantitation (iTRAQ). J Proteome Res 6:821–827
6. Karp NA, Huber W, Sadowski PG et al (2010) Addressing accuracy and precision issues in iTRAQ quantitation. Mol Cell Proteomics 9:1885–1897
7. Kirchner M, Renard BY, Kothe U et al (2010) Computational protein profile similarity screening for quantitative mass spectrometry experiments. Bioinformatics 26:77–83
8. Hultin-Rosenberg L, Forshed J, RMM B et al (2013) Defining, comparing, and improving iTRAQ quantification in mass spectrometry proteomics data. Mol Cell Proteomics 12:2021–2031
9. Vaudel M, Sickmann A, Martens L (2010) Peptide and protein quantification: a map of the minefield. Proteomics 10:650–670
10. Burkhart JM, Vaudel M, Zahedi RP et al (2011) iTRAQ protein quantification: a quality-controlled workflow. Proteomics 11:1125–1134
11. Rauniyar N, Yates JR (2014) Isobaric labeling-based relative quantification in shotgun proteomics. J Proteome Res 13:5293–5309
12. Ow SY, Salim M, Noirel J et al (2009) iTRAQ underestimation in simple and complex mixtures: "the good, the bad and the ugly". J Proteome Res 8:5347–5355
13. Sandberg A, RMM B, Lehtiö J et al (2014) Quantitative accuracy in mass spectrometry based proteomics of complex samples: the impact of labeling and precursor interference. J Proteome 96:133–144
14. Shadforth IP, Dunkley TP, Lilley KS et al (2005) i-Tracker: for quantitative proteomics using iTRAQ. BMC Genomics 6:145
15. Boehm AM, Pütz S, Altenhöfer D et al (2007) Precise protein quantification based on peptide quantification using iTRAQ. BMC Bioinformatics 8:214
16. Vaudel M, Burkhart JM, Radau S et al (2012) Integral quantification accuracy estimation for reporter ion-based quantitative proteomics (iQuARI). J Proteome Res 11: 5072–5080
17. Muth T, Keller D, Puetz SM et al (2010) jTraqX: a free, platform independent tool for isobaric tag quantitation at the protein level. Proteomics 10:1223–1225
18. Arntzen MO, Koehler CJ, Barsnes H et al (2011) IsobariQ: software for isobaric quantitative proteomics using IPTL, iTRAQ, and TMT. J Proteome Res 10:913–920
19. Wen B, Zhou R, Feng Q et al (2014) IQuant: an automated pipeline for quantitative proteomics based upon isobaric tags. Proteomics 14:2280–2285
20. Hu J, Qian J, Borisov O et al (2006) Optimized proteomic analysis of a mouse model of cerebellar dysfunction using amine-specific isobaric tags. Proteomics 6:4321–4334
21. Lin WT, Hung WN, Yian YH et al (2006) Multi-Q: a fully automated tool for multiplexed protein quantitation. J Proteome Res 5:2328–2338
22. Onsongo G, Stone MD, Van Riper SK et al (2010) LTQ-iQuant: a freely available software pipeline for automated and accurate protein quantification of isobaric tagged peptide data from LTQ instruments. Proteomics 10:3533–3538
23. Breitwieser FP, Muller A, Dayon L et al (2011) General statistical modeling of data from protein relative expression isobaric tags. J Proteome Res 10:2758–2766
24. Zhou C, Walker MJ, Williamson AJ et al (2014) A hierarchical statistical modeling approach to analyze proteomic isobaric tag for relative and absolute quantitation data. Bioinformatics 30:549–558
25. Fusaro VA, Mani DR, Mesirov JP et al (2009) Prediction of high-responding peptides for targeted protein assays by mass spectrometry. Nat Biotechnol 27:190–198
26. Hill EG, Schwacke JH, Comte-Walters S et al (2008) A statistical model for iTRAQ data analysis. J Proteome Res 7:3091–3101
27. Fischer M, Renard BY (2016) iPQF: a new peptide-to-protein summarization method using

peptide spectra characteristics to improve protein quantification. Bioinformatics 32:1040–1047

28. Ting L, Rad R, Gygi SP et al (2011) MS3 eliminates ratio distortion in isobaric multiplexed quantitative proteomics. Nat Methods 8: 937–940
29. Carrillo B, Yanofsky C, Laboissiere S et al (2010) Methods for combining peptide intensities to estimate relative protein abundance. Bioinformatics 26:98–103
30. Mahoney DW, Therneau TM, Heppelmann CJ et al (2011) Relative quantification: characterization of bias, variability and fold changes in mass spectrometry data from iTRAQ-labeled peptides. J Proteome Res 10:4325–4333
31. Enke CG (2001) The science of chemical analysis and the technique of mass spectrometry. Int J Mass Spectrom 212:1–11
32. Anderle M, Roy S, Lin H et al (2004) Quantifying reproducibility for differential proteomics: noise analysis for protein liquid chromatography-mass spectrometry of human serum. Bioinformatics 20:3575–3582
33. Gatto L, Lilley KS (2012) MSnbase—an R/Bioconductor package for isobaric tagged mass spectrometry data visualization, processing and quantitation. Bioinformatics 28:288–289
34. Pedrioli PGA, Eng JK, Hubley R et al (2004) A common open representation of mass spectrometry data and its application to proteomics research. Nat Biotechnol 22:1459–1466
35. Martens L, Chambers M, Sturm M et al (2011) mzML—a community standard for mass spectrometry data. Mol Cell Proteomics 10:R110.000133
36. Lai X, Wang L, Tang H et al (2011) A novel alignment method and multiple filters for exclusion of unqualified peptides to enhance label-free quantification using peptide intensity in LC-MS/MS. J Proteome Res 10: 4799–4812
37. Goeminne LJE, Gevaert K, Clement L (2016) Peptide-level robust ridge regression improves estimation, sensitivity, and specificity in data-dependent quantitative label-free shotgun proteomics. Mol Cell Proteomics 15:657–668
38. Anders S, Huber W (2010) Differential expression analysis for sequence count data. Genome Biol 11:R106
39. Robinson MD, McCarthy DJ, Smyth GK (2010) edgeR: a bioconductor package for differential expression analysis of digital gene expression data. Bioinformatics 26:139–140
40. Kammers K, Cole RN, Tiengwe C et al (2015) Detecting significant changes in protein abundance. EuPA Open Proteom 7:11–19
41. Fischer M, Strauch B, Renard BY (2017) Abundance estimation and differential testing on strain level in metagenomics data. Bioinformatics 33:i124–i132
42. Tang H, Arnold RJ, Alves P et al (2006) A computational approach toward label-free protein quantification using predicted peptide detectability. Bioinformatics 22:e481–e488
43. Tsou C-C, Tsai C-F, Tsui Y-H et al (2010) IDEAL-Q, an automated tool for label-free quantitation analysis using an efficient peptide alignment approach and spectral data validation. Mol Cell Proteomics 9:131–144
44. Zhang B, Pirmoradian M, Zubarev R et al (2017) Covariation of peptide abundances accurately reflects protein concentration differences. Mol Cell Proteomics 16:936–948

Chapter 12

Experimental Design in Quantitative Proteomics

Tomasz Burzykowski, Jürgen Claesen, and Dirk Valkenborg

Abstract

Metabolites and proteins are potential biomarkers. They can be identified with the help of mass spectrometry (MS). However, measurements obtained by using MS are prone to various random and systematic errors. The sensitivity of the technology to the errors poses practical challenges, including concerns about reproducibility of the MS-based assays and the possibility of false findings. Given the sensitivity, the proper design of MS-based experiments becomes of utmost importance. In this chapter, we review the basic experimental-design tools that can be used to prevent occurrence of errors that might cause misleading findings in MS-based experiments. We also present results of an experiment aimed at investigating variability of the intensity measurements produced by a MALDI-TOF mass spectrometer. The knowledge about the potential sources of systematic and random errors is fundamental in order to properly design an MS experiment.

Key words Experimental design, Quantitative proteomics, Mass spectrometry, Systematic error, Random error

1 Introduction

Novel "omics" technologies yield a promise for medicine. In particular, they make the concept of precision medicine, i.e., "prevention and treatment strategies that take individual variability into account," [1] more tangible, because they allow researchers to identify the sources of variability in the development and evolution of diseases in a given patient. Toward this aim, *biomarkers* are useful. A biomarker is "a characteristic that is objectively measured and evaluated as an indicator of normal biological processes, pathogenic processes, or pharmacologic responses to a therapeutic intervention" [2]. One can distinguish between *diagnostic biomarkers* that "are used to determine the specific health disorder of the patient" [3], *prognostic biomarkers* that "forecast the likely course of disease irrespective of treatment" [4], and *predictive biomarkers* that "forecast the likely response to a specific treatment" [4].

Caroline A. Evans et al. (eds.), *Mass Spectrometry of Proteins: Methods and Protocols*, Methods in Molecular Biology, vol. 1977, https://doi.org/10.1007/978-1-4939-9232-4_12,

An example of the influence of the availability of biomarkers obtained by using the "omics" technologies is the profound change that has taken place in oncology. Development of molecular-biology-targeted therapies has dramatically changed the range and efficacy of treatments available to cancer patients [5]. At the same time, the use of molecular-biology-based markers improved diagnosis and evaluation of prognosis of the patients. In other disease areas, e.g., cardiovascular conditions [6], similar advancements, boosted by the study of "omics" data, are expected.

Metabolites and proteins are potential biomarkers [7]. They can be identified with the help of mass spectrometry (MS), an analytical technology that is used in proteomics and metabolomics [8, 9]. However, measurements obtained by using MS are prone to various random and systematic errors [10, 11]. The sensitivity of the technology to the errors poses practical challenges, including concerns about reproducibility of the MS-based assays [7]. It can even lead to false findings when the experiment has not been carefully designed.

Perhaps one of the most known examples of experiments using MS, where suboptimal design resulted in false findings, is the study conducted by Petricoin et al. [12]. In the study, SELDI-TOF data were used to train and validate a classifier to discriminate between ovarian cancer samples and control samples. It appeared that a classifier with 100% sensitivity and 95% specificity could be built. In a follow-up paper, using a hybrid quadrupole TOF mass spectrometer, a classifier with 100% sensitivity and specificity could be obtained [13]. However, as it turned out, these results could not be replicated by using an independent dataset [14]. By studying the report and data provided by Conrads et al. [13], Baggerly et al. [14] discovered that, in the experiment, cancer samples were processed on the first day, whereas control samples were processed on the second and third day. However, it appeared that there were issues with the quality of the spectra produced on the third day. Consequently, it could be conjectured that the classifier was actually discriminating between high- and low-quality spectra. However, the latter spectra were present only among the control samples, what resulted in a complete confounding and spurious accuracy of the results of the developed classifier.

In principle, the error in the findings of the experiment could have been avoided if the order of processing of the individual cancer and control samples had been randomized. This example illustrates the importance of experimental design. Given the sensitivity of MS to errors, it is worth to be aware of the experimental design tools that can prevent occurrence of errors that might cause misleading findings. In this chapter, we review the basic tools that are available for this purpose.

2 Basic Principles of Inference

The identification of a diagnostic biomarker requires an evaluation of the association between the biomarker levels and disease status of individuals. The evaluation is most often conducted by using case-control studies, in which samples from persons who are known to be disease free are compared, with respect to biomarker levels, to samples from diseased individuals. This was the approach used in the study by Petricoin et al. [12].

Similarly, for a prognostic biomarker, the association between the biomarker levels at baseline (e.g., at diagnosis or at the start of treatment), or changes of the levels over time, and a clinical endpoint reflecting prognosis (e.g., survival time) has to be established. The association should be independent of treatment. Toward this aim, retrospective analyses of, e.g., clinical trial data can be used.

Finally, to identify a predictive biomarker, it is necessary to establish a dependence of the magnitude of a treatment effect on the levels of a biomarker. Toward this aim, data from randomized clinical trials is usually needed, because reliable and objective estimates of the treatment effect are required.

In each case, we are interested in estimation (and, most likely, testing of the statistical significance) of a parameter that captures the desired association between the biomarker levels and the outcome of interest. For instance, in a diagnostic-biomarker study with a binary biomarker that has got only two levels (low/high, wild-type/mutated, etc.), we may want to estimate sensitivity and specificity that quantify the discriminative properties of the biomarker. These two parameters would also be of interest in a prognostic-biomarker study with a binary biomarker and a binary clinical endpoint (e.g., surviving without disease progression of 3 years). Finally, for a predictive binary biomarker, we are interested in the difference in the treatment effect in the subpopulations of patients defined by the two levels of the biomarker. Note that the treatment effect itself is a parameter (e.g., a difference in mean values) quantifying the difference in a clinical endpoint for patients treated with different treatments.

As in any medical research study, in biomarker studies we draw conclusions based on a limited set (a sample) of individuals representing the population(s) of interest (e.g., healthy and diseased) that are included in the study by some selection mechanism. Given that there are many different samples that can be drawn from the population(s), the estimates of the parameter of interest will necessarily vary, because of the differences between the individuals included in the samples. This variation is a source of error, i.e., the difference between the estimate obtained based on the sample and the true value of the parameter. There are two main types of the error: systematic and random.

Systematic error is often referred to as "bias" and defined as "the expected deviation of an estimate from the true quantity to be estimated" [15]. An example is the so-called selection bias, which occurs when the process of sample selection leads to samples that are not representative for the population or the data-generation mechanism that we study. Another example is the so-called "parallax error," when the user reads an instrument at an angle that results in a readout which is consistently high or consistently low, what results in biased measurements.

Random error may have its source in the heterogeneity among the individuals included in the sample or in the measurement process. Random error does not lead to bias, but implies that the estimates deviate haphazardly, without being systematically too low or too high, from the true quantity to be estimated. In some cases, it is possible to identify some of the sources directly. For instance, we may know that, even if we measure the same sample in a mass spectrometer on two different days, we will obtain somewhat different spectra, but we will not be able to predict the direction of the difference. This knowledge can be used when designing an experiment.

It is worth noting that the concept of bias is not applicable to a single study. In a single study, we obtain an estimate of a parameter which is most likely different from the true value. To claim that the obtained estimate is valid, we should be able to argue that, had we repeated the study many times, we would have obtained unbiased estimates of the parameter, i.e., estimates subject only to the random error. This argumentation can only be based on the properties of the applied experimental design; the collected data themselves cannot provide us with any answer in this respect.

If we can claim that our estimate may be unbiased, an important question is the magnitude of the random error it may be subject to. Experiments in which the expected difference between the estimates and the true value of the parameter is small are said to have a high precision. In statistics, precision is defined as the inverse of the variance of an estimator. Thus, high precision means small variance, i.e., small expected deviation of the estimate from the true value.

3 Basic Principles of Experimental Design

A proper experimental design should aim at two goals: avoiding bias and reducing the magnitude of the random error. Figure 1 illustrates possible situations. In the figure, the center of the target represents the true value of the parameter that we want to estimate. The dots represent the estimates obtained in various studies using some fixed experimental design. Panel A illustrates the situa-

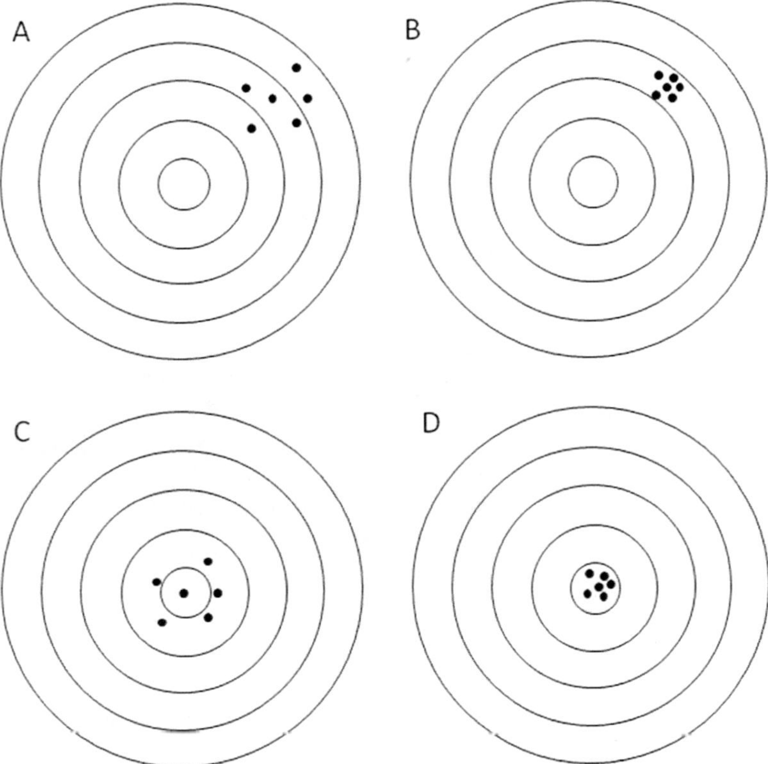

Fig. 1 Bias and precision. A symbolic representation of experimental designs differing with respect to bias and precision. The center of the target represents the true value of the parameter to be estimated. The dots represent the estimates obtained in various studies using the same experimental design. (**a**) Large bias and low precision. (**b**) Large bias and high precision. (**c**) Small bias and low precision. (**d**) Small bias and high precision

tion in which the experimental design will lead to studies providing biased estimates (the dots miss the center of the target) which, in addition, are imprecise (the dots scatter quite a bit around the point they center on). In Panel B, the experimental design results in precise (dots with a low scatter), but biased, estimates. Panel C shows the situation in which the design leads to unbiased estimates (the dots scatter randomly around the center of the target) which lack precision. Finally, Panel D presents the most desirable case of an experimental design that allows unbiased and precise estimation of the parameter.

There are four basic experimental design tools that can be used to avoid bias and influence the magnitude of random error:

1. Control.
2. Randomization.
3. Replication.
4. Blocking (stratification).

The main tools to avoid bias are the use of the control group and randomization. Replication and blocking allow controlling precision, i.e., controlling the magnitude of random error. Blocking also helps in avoiding bias in small studies. We will now discuss each of these techniques in turn.

3.1 Control

Variability in measurements of an outcome that we are interested in may be due to many sources. If, in a study, we want to attribute the differences that we see in the outcome to the levels of a factor (e.g., biomarker) under investigation, we should try to eliminate (control) the potential influence of all other factors on the outcome. For instance, in a comparative study, we should make sure that we include a control group that captures the "natural" mechanisms of the variability. Moreover, in such a study, we should try to make the groups that we compare (e.g., biomarker-negative and biomarker-positive) as similar as possible. If we take the study by Petricoin et al. [12] as an example, we may argue that it involved some (e.g., using the control samples, splitting the data into a training and testing sets), but insufficient, level of control.

3.2 Randomization

In comparative experiments, randomization means a random assignment of the levels of the factor(s), whose effect we are interested in, to the experimental units. A classical example is the random assignment of treatments to patients in clinical trials. In this case, we are interested in the effect of treatment on the health status of patients as measured by a clinical endpoint (e.g., survival time).

Randomization is the primary tool to avoid bias in comparative studies. We will explain its mechanism by using an example of a randomized clinical trial comparing two treatments for cancer, control treatment vs. experimental treatment. Assume that the trial will enroll N patients in total. Moreover, let us assume that we are interested in the effect of the treatment on survival of patients. Consider additionally a patient's characteristic that can influence survival. For instance, in general, patients with more advanced (late-stage) cancer live, on average, shorter than patients with less advanced (early-stage) disease. Assume that, in the trial, due to the "natural" prevalence of different disease stages, a quarter of the patients ($N/4$) will be early stage and three-fourth ($3N/4$) will be late stage.

1. First, suppose that the trial does not use randomization and that the treatment is selected according to the preferences of clinicians and/or patients. If the control treatment is known to show poor efficacy for the late-stage patients, late-stage patients might be more often assigned to the experimental treatment rather than to the control one. As a result, the experimental

treatment group might contain a higher proportion of late-stage patients than the control-treatment group. This, however, could lead to confounding and a biased estimate of the difference in, e.g., survival probability between the two treatments. For instance, even if there was no effect of treatment on survival, the control group might be observed to have a higher probability of survival than the experimental group because of the smaller proportion of the late-stage patients. Or, if the experimental treatment was, in fact, increasing patients' chances of survival, this effect could be "downplayed" by the fact that the treatment was applied to a larger number of late-stage patients. Thus, in such a situation, the estimation of the effect of the experimental treatment would potentially be confounded by patients' stage of disease.

2. Assume now that, as it is typically done in trials similar to the one outlined above, patients are assigned to each of the two treatments with probability 50%. Thus, one can expect that, on average, half of the patients ($N/2$) will be assigned to the control treatment, with the other half getting the experimental therapy.

Moreover, if randomization is properly conducted, there should be a 50–50 split of early-stage patients between the control and experimental treatments. This means that the number of such patients getting the control treatment should be about $(N/4)/2 = N/8$, with (approximately) the same number getting the experimental treatment. A similar argument could be applied to late-stage patients, with $(3N/4)/2 = 3N/8$ getting the control and the experimental treatment.

From those calculations it follows that the proportion of early-stage patients in the control group should approximately be equal to $(N/8)/(N/8 + 3N/8) = 25\%$, with the same proportion applying to the experimental group. This means that the two treatment groups should be comparable with respect to the distribution of the stage of disease. Hence, there should be no confounding of the effect of the experimental treatment by stage of disease. For instance, if there was no effect of treatment on survival, both the control group and the experimental group should be observed to enjoy a similar survival probability.

One could argue that, in the case of confounding as described above, one could remove bias in the estimate of the effect of treatment by using an appropriate method of statistical analysis. In particular, the method would have to take into account the difference in the distribution of the stage of disease in the two treatment groups. However, such an analysis can only be conducted if the confounding factor is known to us. Obviously, this is not always the case. In this respect it is worth noting that the "balancing" mechanism of randomization applies to all factors, whether known or unknown. The avoidance of bias due to confounding by unknown factors is actually the most important feature of randomization.

The erroneous findings in the Petricoin et al. [12] study were due to occurrence of confounding between the day of processing of the sample and the status of the sample. Note that the influence of the day of experiment on the (quality of) MS measurements occurred by chance. However, because of the chosen order of processing of the samples (cancer samples processed on day one, control samples processed on day two and three), the day distribution became very much imbalanced between the cancer and control groups. In fact, the proportion of the day-three (lower-quality) spectra in the cancer group was equal to 0, while it was non-zero in the control group. As a result, day became a confounding factor.

The confounding effect could have been prevented by a simple randomization of the order of processing of samples. As a result, one would expect to see some of the cancer samples processed on day three. Consequently, the lower-quality spectra would have also appeared in the cancer group and, in this respect, the groups would have become comparable.

3.3 Replication

Replication means the repetition of experimental conditions to multiple independent experimental units. For instance, in a randomized clinical trial, each of the compared treatments is applied to multiple patients. Each patient is thus a *replicate*.

The main goal of replication is to reduce the magnitude of the random error. Assume that we are interested in estimating a mean value of a variable. Let the mean value be equal to μ. Assume that the standard deviation of the variable is equal to σ. This means that a single observation of the variable will differ from μ, on average, by about σ. Equivalently, we can say that standard deviation of the random error present in a single observation is equal to σ.

On the other hand, if we collect a sample of N independent observations of the variable, then the sample mean of these observations will differ from μ, on average, by about $\sigma / \sqrt{N}$. Equivalently, we can say that standard deviation of the random error in the sample mean is equal to $\sigma / \sqrt{N}$. This implies, however, that the precision of the sample mean as an estimator of μ increases N times. (Precision is the inverse of the variance, so for a single observation precision is equal to $1/\sigma^2$, while for the sample of N observations precision of the sample mean is equal to N/σ^2; hence, the N-fold increase in precision).

One should distinguish between independent replicates and repeated measurements. For instance, if, in an experiment, a treatment is applied to four patients and then a clinical endpoint (e.g., blood pressure) is measured for each patient, we have got four replicates. However, if the treatment is applied to a single patient, and then the clinical endpoint is measured for this patient four times, we have got four *repeated measurements* rather than four replicates. The important difference is that the four measurements obtained for a single patient may be correlated, and their variation is most likely different from variation of four (independent) measurements obtained for four different patients. In fact, the four repeated measurements reflect only the (technical) variability related to the measurement process of the clinical endpoint, but they do not reflect the between-patient (biological) variability of the endpoint. By increasing the number of repeated measurements of the single patient, we could reduce the random error related to the measurement process to a minimum. This would, however, only bring us to the knowledge about the clinical endpoint for the single patient, without any possibility to generalize the results of the experiment beyond that patient. To do so, we need independent replicates.

It is possible to calculate the number of replicates (the *sample size*) necessary to reach the desired level of precision. Toward this end, however, some information about at least standard deviation of the random error is required. Often, this information may require running a small "pilot" experiment, in which a few replicates are used to get an estimate of the standard deviation.

3.4 Blocking

Blocking is a technique aimed, primarily, at increasing precision in comparative experiments. It is based on the intuitive idea of comparing "like with like" [16]. To implement blocking, we require prior information about factors that may be related to the outcome we are interested in, i.e., potential confounding factors. Subsequently, the experimental units are divided into groups (*blocks*) that share the same levels of the potential confounding factors. We then compare the differences in the outcome between the units within each block that differ by the levels of the factor(s) we are primarily interested in.

For instance, a classical example of an experiment with blocking is a "cross-over trial." In the simplest case when the trial compares two treatments, the treatments in a cross-over trial are administered sequentially (in a random sequence) to the same patient. Subsequently, the effect of the treatment is estimated by comparing the values of a clinical outcome obtained under the two treatments for each individual patient (by, e.g., taking the within-patient differences). This way, the effect can be estimated more precisely than by comparing the values of a clinical outcome for different patients treated with different treatments. This is because, by using the within-patient comparison, we remove the between-patient variability from the estimation.

Blocking can also help in preventing bias due to confounding. As mentioned earlier, randomization should in principle be able to remove the effect of confounding variables in comparative studies. Sometimes, however, randomization may accidentally fail to balance exactly the distribution of the levels of confounding factor(s) between the compared groups. The chance of coming across such a situation increases especially for experiments with a small number of experimental units and for an increasing number of confounding factors. In such case, if we know the potential confounding factor(s), we can use blocking to ensure that the distribution of the levels of the factor(s) is balanced between the groups we want to compare.

Let us explain the technique by using the example of a randomized clinical trial comparing the standard and experimental treatment for cancer with disease stage as a potential confounding factor. In that case, in order to balance the distribution of early-stage and late-stage patients in the control and experimental groups, randomization should be conducted separately for the early-stage and late-stage patients, i.e., within the two *strata* defined by the two possible levels of the disease-stage factor. These strata form, effectively, the blocks. Within each stratum, we may additionally conduct the randomization in such a way that the number of patients assigned to each of the two treatments is (approximately) the same. Toward this aim, we could define groups of, for instance, four consecutive patients to be enrolled in a particular stratum. In each group, the randomization could be conducted in such a way that two patients are assigned to the control treatment and two patients are assigned to the experimental treatment. In particular, one could consider selecting (randomly) one of the following assignment sequences for each group: CCEE, CECE, CEEC, ECCE, ECEC, EECC (where C stands for the control and E for the experimental treatment). This implies that, after enrolling four consecutive patients, the number of patients assigned to each of the two treatments in a particular stratum would be equal. Consequently, the proportion of early-stage and late-stage patients in each treatment group would be approximately the same.

This technique of *stratified randomization* can be applied to more than one confounding factor. In that case, randomization should be conducted within the strata defined by the combinations of the levels of the factors. However, the number of strata should not be too large, because one may not be able to complete randomization in all the blocks (due to the insufficient number of patients within some strata). In that case, the balancing effect of stratified randomization will not be achieved.

It is also possible to apply the technique to factors that are known to be a source of random error. In small experiments such factors can also become a source of bias if the distribution of their levels between the compared groups is imbalanced and the factors are related to a large variability.

In the study of Petricoin et al. [12], one could have tried to eliminate the confounding effect of the day of sample processing by applying blocking. In particular, one could have considered each day as a block. Then, within each block (day), one would select randomly half of the samples that could be processed on that day from the control group and the other half from the cancer group. As a result, the same proportion of the day-three (lower-quality) spectra in both groups would be obtained. Of course, this would require recognition of the day as a potential confounding factor. However, as discussed in the next section, it appears that day effects are an important source of random variability in MS-based experiments. Hence, it is always worth considering day as a potential confounding factor while designing such experiments.

4 Variability of MALDI-TOF Measurements

From the discussion of the experimental-design tools it should be clear that, in order to properly design an experiment, we should have an idea about the potential sources of systematic and random error. Thus, regardless of the technological platform, it is important to conduct experiments aimed at getting knowledge about the potential sources of variability of the measurements obtained by using the platform.

In this section, we describe an experiment that was run to investigate variability in the intensity measurements produced by a MALDI-TOF mass spectrometer. (The details regarding the particular equipment used in the experiment cannot be discussed due to confidentiality reasons, but they are irrelevant for our purposes.) In particular, eight stainless-steel plates with 640 spots arranged in a 20×32 grid were used. The same sample, with a mixture of eight known peptides, was spotted on all plates. Five potential sources of variability in intensity measurements were included:

1. Plate.
2. Day: plates 1–4 were processed on day 1, plates 5–8 were processed on day 2.
3. Volume of the sample per spot: 0.25 (on plates 1, 4, 5, 8) and 0.50 (on plates 2, 3, 6, 7).
4. Number of times each spot was shot by the laser (shooting pattern): 25 series of 40 pulses (25×40), 25×80, 40×50, 50×60.
5. Shooting method: centrally at the spot, spiral, at the edges of the spot, uniform.

Table 1
Proposed experimental design

C	E	S	U	*C*	*E*	*S*	*U*	C	E	S	U	**C**	**E**	**S**	**U**	...
S	U	C	E	*S*	*U*	*C*	*E*	S	U	C	E	**S**	**U**	**C**	**E**	...
E	C	U	S	*E*	*C*	*U*	*S*	E	C	U	S	**E**	**C**	**U**	**S**	...
U	S	E	C	*U*	*S*	*E*	*C*	U	S	E	C	**U**	**S**	**E**	**C**	...
C	**E**	**S**	**U**	C	E	S	U	*C*	*E*	*S*	*U*	C	E	S	U	...
S	**U**	**C**	**E**	S	U	C	E	*S*	*U*	*C*	*E*	S	U	C	E	...
E	**C**	**U**	**S**	E	C	U	S	*E*	*C*	*U*	*S*	E	C	U	S	...
U	**S**	**E**	**C**	U	S	E	C	*U*	*S*	*E*	*C*	U	S	E	C	...
C	E	S	U	**C**	**E**	**S**	**U**	C	E	S	U	*C*	*E*	*S*	*U*	...
S	U	C	E	**S**	**U**	**C**	**E**	S	U	C	E	*S*	*U*	*C*	*E*	...
E	C	U	S	**E**	**C**	**U**	**S**	E	C	U	S	*E*	*C*	*U*	*S*	...
U	S	E	C	**U**	**S**	**E**	**C**	U	S	E	C	*U*	*S*	*E*	*C*	...
C	*E*	*S*	*U*	C	E	S	U	**C**	**E**	**S**	**U**	C	E	S	U	...
S	*U*	*C*	*E*	S	U	C	E	**S**	**U**	**C**	**E**	S	U	C	E	...
E	*C*	*U*	*S*	E	C	U	S	**E**	**C**	**U**	**S**	E	C	U	S	...
U	*S*	*E*	*C*	U	S	E	C	**U**	**S**	**E**	**C**	U	S	E	C	...
C	E	S	U	*C*	*E*	*S*	*U*	C	E	S	U	**C**	**E**	**S**	**U**	...
S	U	C	E	*S*	*U*	*C*	*E*	S	U	C	E	**S**	**U**	**C**	**E**	...
E	C	U	S	*E*	*C*	*U*	*S*	E	C	U	S	**E**	**C**	**U**	**S**	...
U	S	E	C	*U*	*S*	*E*	*C*	U	S	E	C	**U**	**S**	**E**	**C**	...

The table represents the arrangement of the shooting methods and shooting patterns across the spots on a plate. Shooting methods: C (central), E (edges), S (spiral), U (uniform). Shooting pattern: 25 × 80 (normal font), 25 × 40 (italic), 40 × 50 (underscore), 50 × 60 (bold).

The arrangement of the shooting methods and shooting patterns across the spots on a plate is shown in Table 1. In essence, the experiment treated columns and rows of a plate as "blocks." In particular, a Latin-square design was used for the shooting patterns, with patterns cycled across all 32 columns and 20 rows of a plate. Each shooting pattern was applied to an array of 4 × 4 spots. Within the array, the four shooting methods were applied in a Latin square. By using the design presented in Table 1, each shooting pattern appeared exactly two times in each row and at least once in each column. Also, each shooting method appeared exactly eight times in each row and five times in each column.

The design yielded, in total, 640 × 8 = 5,120 spectra. For each spectrum, the logarithms of total-ion-count (TIC) adjusted intensity measurements for each of the peptides were obtained.

The analysis was conducted separately for each of the peptides. Initially, an ANOVA model with shooting pattern, shooting method, their interaction, and volume as the fixed effects, and day, plate, and spot as the random effects, was used to estimate the variability associated with each of the factors. Subsequently, the model was modified by removing the volume and spot effects, and assuming that residual variance depended on shooting pattern and method.

The effect of volume was found not to be statistically significant (by applying the two-sided significance level of 0.05) for any of the peptides in the initial ANOVA model. There was a statistically significant effect of the combination of the shooting pattern and shooting method: the mean values of the logarithms of TIC-adjusted intensity measurements for each of the peptides were, in general, the highest for the spiral shooting method and the smallest for the edge-concentrated method (Fig. 2). However, the differences were not large.

Figure 3 presents the estimated variances related to the day, plate, and spot effects, as well as the residual (unknown-source random-error) variance (assumed constant) based on the initial ANOVA model. The figure suggests a substantial effect of day: the between-day variability is larger than the between-plate variability for six out of the eight analyzed peptides. The between-plate variability is, in turn, larger than the residual variability (i.e., between-spot variability within each plate). The across-plate spot-related variability is negligible.

Figure 4 presents the estimated residual variances for various combinations of the shooting method and pattern obtained from the ANOVA model that assumed the dependence of the residual variance on the pattern and method. The variability of the measurements seems to be the smallest for the "uniform" method (with the exception of the two heaviest peptides), perhaps combined with 40 × 50 or 50 × 60 shooting patterns (as seen, e.g., for peptides with masses 1168 m/z, 1434 m/z, or 1456 m/z). For this method, the variance remains below 0.1 for the six lightest peptides, i.e., is smaller than the variance due to the day and plate effects (Fig. 3).

These findings suggest that, when designing a comparative experiment using the particular MALDI-TOF technology, one should consider day and plate as potential confounding effects. Hence, day and plate should be used in blocking, and randomly selected samples from the compared groups should be best compared on the same plate. Spot effect can be ignored, i.e., samples can be assigned haphazardly to the spots. Volume effect can, in principle, also be ignored, but it might be worthwhile to try to spot approximately the same amount of material on each spot. Finally, it may be more advantageous to generate spectra by using the "uniform" shooting method combined with 40 × 50 or 50 × 60 shooting patterns.

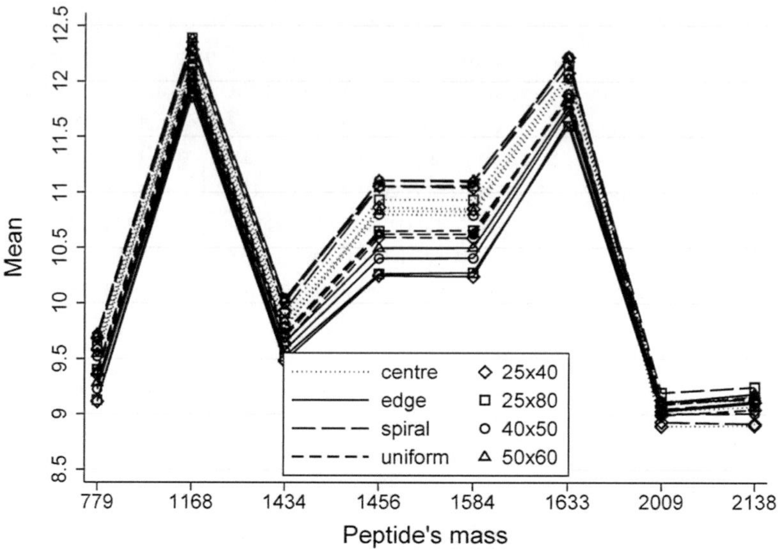

Fig. 2 Peptide-specific intensity and effect of shooting patterns and methods. Model-based estimates of the peptide-specific mean values of the logarithms of TIC-adjusted intensity measurements for different combinations of the shooting patterns (indicated by the line patterns) and shooting methods (indicated by the symbols) as a function of the peptide mass

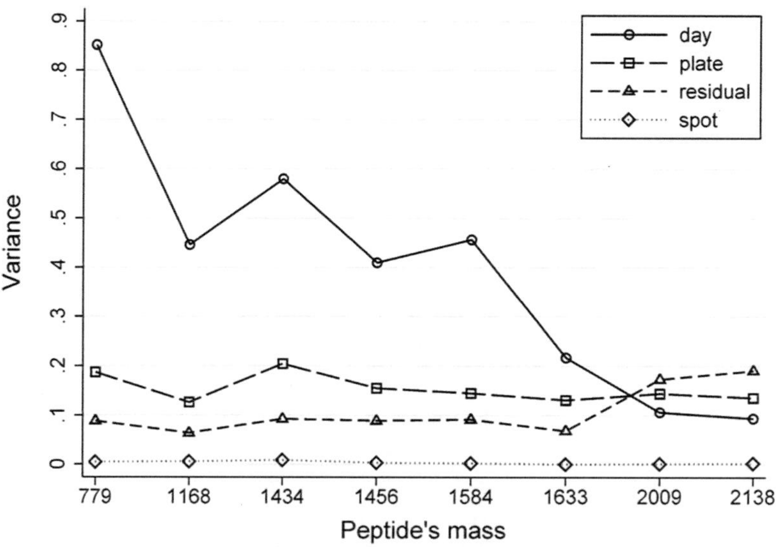

Fig. 3 Variances for day, plate, and spot effects. Model-based estimates of peptide-specific variances related to the day, plate, and spot effects, as well as the residual (unknown-source random-error) variance (assumed constant) as a function of the peptide mass

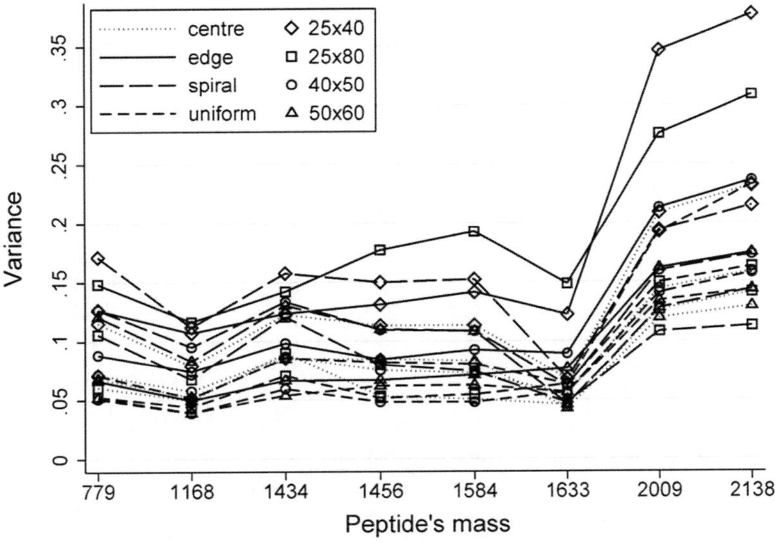

Fig. 4 Residual variance for shooting patterns and methods. Model-based estimates of peptide-specific residual variances for different combinations of the shooting patterns (indicated by the line patterns) and shooting methods (indicated by the symbols) as a function of the peptide mass

5 Statistical Modeling

Here we provide the lines of R code that allow for the statistical modeling described in this chapter.

Assume that the data are stored using the "long" format in data frame "all":

Day	Plate	Spot	Patt	Meth	Vol	Peptmeas
1	1	A1	25_80	c	2	10.3007
1	1	A2	25_80	s	2	9.83189
1	2	A1	50_60	u	1	10.0769
1	2	A2	50_60	e	1	10.4503
1	3	A1	40_50	e	1	10.3803
1	3	A2	40_50	u	1	9.69652
1	4	A1	25_40	s	2	10.5655
1	4	A2	25_40	u	2	9.1195
2	1	A1	25_80	c	2	9.44061
2	1	A2	25_80	s	2	8.43269
2	2	A1	50_60	u	1	10.1359
2	2	A2	50_60	e	1	9.38974

Day	Plate	Spot	Patt	Meth	Vol	Peptmeas
2	3	A1	40_50	e	1	10.2826
2	3	A2	40_50	u	1	9.00234
2	4	A1	25_40	s	2	10.1796
2	4	A2	25_40	u	2	8.57258

The model underlying the results presented in Fig. 3 can be fitted by using function *lmer()* from the R package *lme4*:

```
library(lme4)

mod1 <- lmer(peptmeas ~ vol + patt * meth +
 (1 | day) + (1 | plate:day) + (1 | spot), data=all)
summary(mod1)
detach(package:lme4)
```

Note that effect of the random spot effect is crossed with the plate (nested within the day) effect, i.e., assumed to be the same across plates. This assumption can be argued with, but it allows checking the magnitude of the between-plate variability of spot-specific measurements that may be due to, for instance, systematic effects of some spots (e.g., located at the plate edges). The within-plate between-spot variability is captured by the residual variance (assumed constant).

The model underlying the results presented in Fig. 4 can be fitted by using function *lme()* from the R package *nlme*:

```
library(nlme)

mod2 <- lme(peptmeas ~ -1 + patt:meth,
     random = list(day=~1, plate =~1),
     weights=varIdent(form = ~ 1 | patt*meth), data=all)
summary(mod2)
detach(package:nlme)
```

For the spots, this model assumes only within-plate variability that is captured by the residual variance. The variance is assumed to depend on the applied combination of shooting pattern and method.

6 Conclusion

Modern “omics” technologies are complex and, hence, vulnerable to systematic and random errors. To avoid false findings when using the technologies, one should identify the potential sources of the errors. The knowledge should then be used when planning experiments in order to avoid bias and increase precision of the inference based on the experimental results. All these comments apply to proteomic and metabolomic studies conducted with the use of MS.

References

1. Collins FS, Varmus H (2015) A new initiative on precision medicine. N Engl J Med 372:793–795
2. Biomarkers Definitions Working Group (2001) Biomarkers and surrogate endpoints: preferred definitions and conceptual framework. Clin Pharmacol Ther 69:89–95
3. Quezada H, Guzmán-Ortiz AL, Díaz-Sánchez H et al (2017) Omics-based biomarkers: current status and potential use in the clinic. Bol Med Hosp Infant Mex 74:219–226
4. Buyse M, Sargent DJ, Grothey A et al (2010) Biomarkers and surrogate end points—the challenge of statistical validation. Nat Rev Clin Oncol 7:309–317
5. Roychowdhury S, Chinnaiyan AM (2016) Translating cancer genomes and transcriptomes for precision oncology. CA Cancer J Clin 66:75–88
6. Shehadeh LA, Hare JM (2013) Ribonucleic acid biomarkers for heart failure: is there a correlation between heart and blood transcriptomics? JACC Heart Fail 1:477–479
7. Hathout Y (2015) Proteomic methods for biomarker discovery and validation. Are we there yet? Expert Rev of Proteomics 12:329–331
8. Eidhammer I, Flikka K, Martens L, Mikalsen S-O (2007) Computational methods for mass spectrometry proteomics. John Wiley & Sons, Chichester
9. Datta S, Mertens BJA (eds) (2016) Statistical analysis of proteomics, metabolomics, and lipidomics data using mass spectrometry. Springer, New York
10. Hilario M, Kalousis A, Pellegrini C, Muller M (2006) Processing and classification of protein mass spectra. Mass Spectrom Rev 25:409–449
11. Ejigu BA, Valkenborg D, Baggerman G et al (2013) Evaluation of normalization methods to pave the way towards large-scale LC-MS-based metabolomic profiling experiments. OMICS 17:473–485
12. Petricoin EF III, Ardekani AM, Hitt BA et al (2002) Use of proteomic patterns in serum to identify ovarian cancer. Lancet 359:572–577
13. Conrads TP, Fusaro VA, Ross S et al (2004) High-resolution serum proteomic features for ovarian cancer detection. Endocr Relat Cancer 11:163–178
14. Baggerly KA, Morris JS, Coombes KR (2004) Reproducibility of SELDI-TOF protein patterns in serum: comparing datasets from different experiments. Bioinformatics 20:777–785
15. Gail MH (2005) Bias. In: Encyclopedia of biostatistics. Wiley, New York
16. Cox DR, Reid N (2000) The theory of the design of experiments. Chapman & Hall/CRC, Boca Raton

Chapter 13

Practical Integration of Multi-Run iTRAQ Data

Dana Pascovici, Xiaomin Song, Jemma Wu, Thiri Zaw, and Mark Molloy

Abstract

In this chapter, we describe some of the approaches we employ in the analysis of iTRAQ data in our group, with an emphasis on practical issues that can occur in larger multi-run projects. Our pipeline starts with a well-established iTRAQ workflow, makes use of protein level quantitation using ProteinPilot, and continues either via a global analysis in the presence of a common reference, or by identifying pairwise comparisons of interest and applying a method taking the protein ratios and protein ratio confidence measures into consideration. Additionally we describe what issues can occur in the more subtle scenarios involving composite databases in multi-run situations, and an approach applicable in that setting.

Key words Mass spectrometry, Quantitative proteomics, iTRAQ, Data processing, Replication

1 Introduction

Since its development and introduction a decade and a half ago, the popularity of shotgun proteomics using isobaric labeled tags has not abated. iTRAQ [1] experiments are carried out in many laboratories alongside TMT [2], counts and intensity-based label-free approaches, and more recent label-free approaches such as DIA/SWATH workflows [3]. The ability to multiplex samples in a single run and thus both remove experimental variation between conditions and decrease instrument time usage remains attractive to researchers. Over the last decade, the focus has shifted to biological comparisons, with increasingly complex scenarios becoming possible. This trend has been aided by the larger panels of isobaric tags available in the 8-plex iTRAQ or 10-plex TMT, thus allowing for more samples to be placed on the same run, with the possibility of even larger panels of 16 to 26-plex becoming available in the future [4].

To the extent that an experiment fits on a single iTRAQ or TMT run, its analysis can be straightforward, and the advantages of thus removing experimental variability are dominant. Software suites such as ProteinPilot (SCIEX), Proteome Discoverer

Caroline A. Evans et al. (eds.), *Mass Spectrometry of Proteins: Methods and Protocols*, Methods in Molecular Biology, vol. 1977, https://doi.org/10.1007/978-1-4939-9232-4_13,

(Thermo), Mascot (Matrix Science), or Pro QUANT (ABI) can be employed, or tools developed by various proteomics groups [5–7]. Most software suites will provide protein and peptide relative quantitation, including measures of confidence such as confidence intervals, *p*-values, or coverage percentages, and offering rapid estimation of identification false discovery rates. Many software suites report protein grouping for sets of proteins identified by shared peptides, and thus correctly provide realistic, parsimonious sets of protein identifications [8].

There are however several well-known bioinformatics challenges related to protein and peptide isobaric quantitation determination which have been well summarized in a recent review [9], a core issue being that of co-eluting peptides making estimated ratios appear compressed, and variably so depending on the complexity of the background [10, 11]. Other known issues are the impact of the fraction on the peptide ratio estimation (fraction effect), the variability of protein ratio estimation depending on the underlying peptides (peptide effect), and the fact that multiple peptides increase the confidence of the quantitation. These aspects, blended with others such as missing data and potential technical variability across runs, can make the analysis of a large experiment spanning multiple iTRAQ runs a difficult task, as working with the protein level quantitation only may be too simplistic, but working with the lower level peptide/ion information can quickly become computationally complex.

In addition, increasingly analyses are undertaken with non-model species or plants where the complexity of the database increases [12], or in multi-species or multi-strain situations [13, 14], experiments that by their nature contain many highly similar proteins from related organisms, thus containing many shared peptides. With a complex database used for identification the consistency of identifications can potentially decrease in multi-run scenarios [15], due to the stochastic nature of assigning a winner to a protein group—both the protein groups and the assigned group winners can vary across runs [16], thus adding layers of complexity to the analysis and interpretation of data.

Thus, the need remains for reliable and wherever possible easy to understand and interpret methods of data analysis of iTRAQ data. Statistically sophisticated methods are available [17, 18] and correctly account for the variability at all the lower levels (ion, peptide, protein), albeit at the expense of increased complexity. There is also a place for trialling simpler, pragmatic methods of analysis that may be more easily approached and interpreted. The important starting points are high-quality peptide and protein quantitation data, well-established protocols for both the iTRAQ workflow and the subsequent data analysis, and awareness of the possible pitfalls arising particularly in the presence of numerous runs or complex databases.

In this chapter, we describe some of the pragmatic approaches used in our laboratory for data analysis of multi-run iTRAQ experiments. The starting data in our group is the protein quantitation as provided by Protein Pilot, with the analysis aim being to provide differentially expressed proteins between pairs of conditions identified as of interest by the researcher. The quantitation data from different runs is combined, and the pairwise comparisons of interest are identified. The differentially expressed proteins are identified based on their combined Stouffer *p*-value [19], which combines individual protein ratio *p*-values in the style of a meta-analysis, and trend consistency measure, as described in detail in our previous work [20]. We also describe the fill-down quantitation option that aims to address the problems arising out of potential protein group winner mismatches in the case of redundant, multi-species databases.

2 Materials

The materials needed for running an iTRAQ experiment are not detailed in this chapter. More comprehensive example data, R code, and results are included in the references describing our methodology [15, 19].

1. ProteinPilot or other software generating iTRAQ protein ratios and *p*-values.
2. GenePattern can be downloaded from http://software.broadinstitute.org/cancer/software/genepattern/download as a server to provide internally a web interface for R scripts.
3. The R programming environment can be downloaded from https://cran.r-project.org/
4. Set of custom R scripts (iTRAQRpac).

3 Methods

The bioinformatics workflow detailed below uses iTRAQ and the protein quantitation results provided by Protein Pilot, but could be employed more broadly to be used wherever protein ratios and a measure of significance determined from the underlying peptides are available. The iTRAQ methodology employed has been described previously in detail [21–23].

3.1 Isobaric Labeled Proteomics

The dataset used for the examples in this chapter is extracted from a publicly available experiment on a wheat pathogen exposure time course [20, 21]. The experiment analyzed wheat leaves infiltrated with either a wheat toxin or an empty vector over four time points, with three biological replicates (pools of plants grown in the same pot) at each time point. The labeled samples were placed in eight 4-plex runs. The biological details are not the focus of this chapter; hence, for illustration purposes we have selected a subset of the runs and a range of pairwise comparisons where the changes were subtle (few proteins expected to be differentially expressed, small fold changes, no visible changes on the leaf), and a number of comparisons where the changes were strong (many differentially expressed proteins, large fold changes, visible changes in the leaf).

When a global analysis is of interest, introducing a common reference (such as a pool or repeated sample) is often preferred due to its simplicity [9] but is not required; including a reference in all runs may not be a burden when an 8-plex is being used, but can be a considerable burden with 4-plex, as it occupies 25% of the channels. The advantage of having a global reference is that all protein ratios can be extracted with respect to that reference, and a standard multivariate analysis (PCA, clustering, ANOVA) can be undertaken on log-transformed protein ratios to give a rapid overview of the data.

However, in a variety of experiment types, of interest may be only targeted pairwise comparisons, for instance, across a stress/no stress time course [21], in paired comparisons of affected/non-affected tissue from the same patients, or in situations where experiments are repeated several times and combined [13, 24]. In such situations, the comparisons of interest are often placed on the same run, and are then combined across runs. This is the scenario we focus on here.

The analysis described is carried out using a series of R scripts made available internally via a GenePattern server but which can be made available upon request. Source R code covering most of these approaches was also made available as described in our publications [16, 20].

1. Experiment design. As within-run technique variations are normally smaller than across runs, including samples to be compared within a run may help identify smaller differences between samples. For an iTRAQ multiple run experiment including biological and technical replicates, a common sample in all runs can enable individual run data quality checking and overall ANOVA analyses. However, this is not necessary for pairwise sample comparisons.
2. Extract proteins. The method for protein extraction depends on the biological sample. Samples in one iTRAQ experiment should be treated with the same protein extraction buffer and the same procedures to minimize sample preparation variation. Primary amines are not compatible with iTRAQ labeling. If chemicals with primary amines are used in protein extraction, protein samples need to be cleaned up with acetone precipitation.

3. Digest protein samples with trypsin. This can be done by reducing protein samples with 5 mM Tris(2-carboxyethyl) phosphine (TCEP) for 1 h at 60 °C, and alkylating with 10 mM *S*-ethylmethanethiosulfonate (MMTS) at room temperature for 10 min; digesting a total of 100 μg of protein overnight with trypsin at 37 °C at a ratio of 1:20 trypsin to protein.
4. Label digested samples with the iTRAQ reagents following the protocol provided by the vendor (SCIEX). One vial of iTRAQ reagent is used for every 100 μg of protein. Equal amount of proteins should be labeled for all iTRAQ tags in each run. Ethanol is used to solubilize the iTRAQ reagent then added to the peptide sample ensuring a final organic concentration of at least 60% (v/v). After 1 h of iTRAQ labeling, the reaction is quenched by adding 50 μL of water. Then the samples are mixed.
5. If samples are complex, fractionate iTRAQ labeled peptides by strong cation exchange liquid chromatography. Otherwise, one can acquire LC ESI MS/MS data for the labeled samples without fractionation.
6. Acquire LC ESI MS/MS data for pooled iTRAQ labeled samples using positive ion electrospray with a tandem mass spectrometer such as TripleTOF 5600 (SCIEX) in an information-dependent acquisition mode (IDA).

3.2 Identify and Quantify Proteins

1. Identify and quantify proteins for each iTRAQ run data using mass spectrometer vendor specific database searching software. For example, use ProteinPilot to process data acquired with a SCIEX TripleTOF mass spectrometer.
2. Export the protein summary file as a tab-delimited .txt file. Make sure all protein ratios with respect to all references are exported for future analysis.
3. Filter protein summary file to only keep proteins with confidence higher than 95%.

3.3 Pairwise Comparisons

The iTRAQ run design is tightly linked with the data analysis as envisaged at the start of the experiment, as analysis plans are usually taken into consideration when determining how to place samples on runs. Then when starting the data analysis, the starting point is examining the design and considering possible analysis options. In the context of this book chapter, we focus on the pairwise comparisons, as the overall methodology employed is commonly used and well understood. We describe the steps in the context of our example dataset.

The possible data analysis workflow options are presented in Fig. 1, and an example experiment design is presented in Fig. 2.

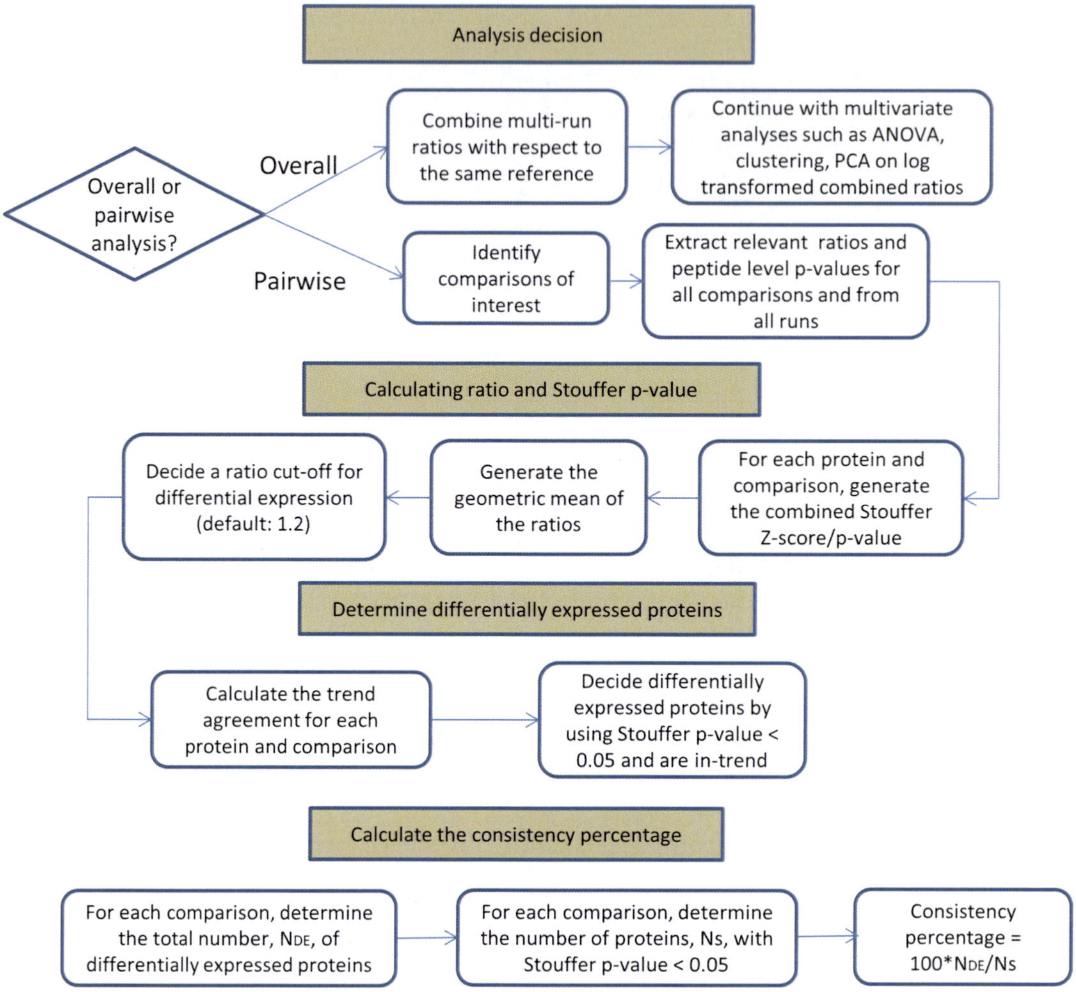

Analysis decision

Overall or pairwise analysis?

Overall

Combine multi-run ratios with respect to the same reference

Continue with multivariate analyses such as ANOVA, clustering, PCA on log transformed combined ratios

Pairwise

Identify comparisons of interest

Extract relevant ratios and peptide level p-values for all comparisons and from all runs

Calculating ratio and Stouffer p-value

For each protein and comparison, generate the combined Stouffer Z-score/p-value

Generate the geometric mean of the ratios

Decide a ratio cut-off for differential expression (default: 1.2)

Determine differentially expressed proteins

Calculate the trend agreement for each protein and comparison

Decide differentially expressed proteins by using Stouffer p-value < 0.05 and are in-trend

Calculate the consistency percentage

For each comparison, determine the total number, N_{DE}, of differentially expressed proteins

For each comparison, determine the number of proteins, Ns, with Stouffer p-value < 0.05

Consistency percentage = $100*N_{DE}/Ns$

Fig. 1 Potential data analysis options workflow in a multi-run experiment

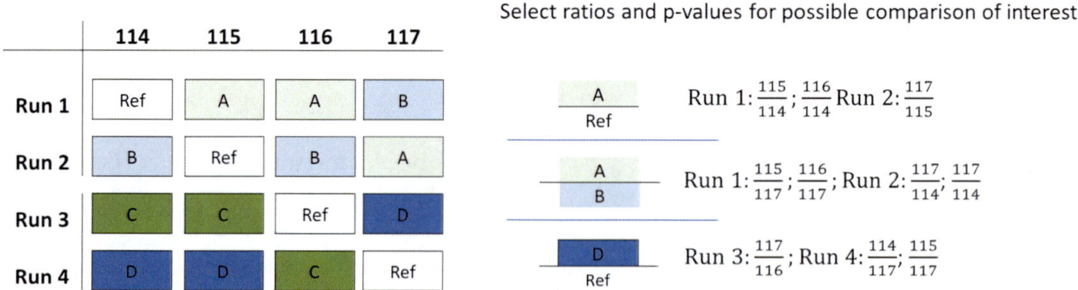

Fig. 2 Example of selecting ratios and peptide level *p*-values for some potential comparisons of interest

1. If in the experiment at hand a common reference is present, the protein ratios with respect to that ratio can be extracted from all runs and then can be combined and analyzed using approaches such as ANOVA, PCA, or clustering to identify differentially expressed proteins. This approach is easy to implement, though possibly simplistic as high and low confidence ratios are now treated in the same way; also the approach assumes no variability in the pool across runs. The peptide level *p*-values are not used in this approach, though additional filters based for instance on the number of peptides could be used at the end.
2. Alternatively, it is possible that only several pairwise comparisons are of interest. Figure 2 shows some possible example comparisons in the context of this experimental design.
3. Identify all comparisons of interest for the experiment at hand; the remaining workflow focuses on this pairwise comparison scenario.
4. For each comparison of interest identify all the relevant ratios from each individual runs. In Fig. 2, we identify a few comparisons of interest: one where subtle changes with no visible phenotype changes are present (A/Ref or B/Ref), and one where strong changes with visible phenotype changes are present (D/Ref or C/Ref). Many other comparisons are possible.
5. Extract all the relevant ratios for all the comparisons and all the runs, along with their respective peptide *p*-values. An example subset is tabulated for the two comparisons of interest in Table 1.

3.4 Stouffer *p*-Value and Trend Agreement

1. For each protein and each comparison, generate the combined Stouffer *Z*-score from the peptide level *p*-values of all ratios being combined. First the *p*-values are transformed into *Z*-scores using the normal quantile function (e.g., -qnorm(p) in R). Then the *Z*-scores are combined, $Z = \frac{\sum_1^k Z_i}{\sqrt{k}}$, where k is the number of scores being combined and Z_i is the *Z*-score of sample i, and finally the result is translated back to a *p*-value using the normal distribution function (e.g., pnorm(Z, lower.tail=FALSE) in R). Several examples are shown in Table 1, ranked in order of increasing Stouffer *p*-value. Note that proteins with very low Stouffer *p*-values typically are identified by many peptides.
2. For each protein and each comparison, generate the geometric mean of the ratios—an example is shown in Table 1, column Geomean (*see* **Note 1**).

Table 1
Example Stouffer combined *p*-value and trend value calculation

B/Ref	Accession	#Peptides	Run 1 117/114	*p*-Value 117/114	Crit	Run 2 114/115	*p*-Value 114/115	Crit	Run 2 116/115	*p*-Value 116/115	Crit	Geomean	Stouffers *p*-value	Trend value	*t*-Test *p*-value
Subtle effect	Q1XIR9_WHEAT	43	0.91	0.1033	0	0.66	0.0006	−1	1.21	0.0004	1	0.90	$2.944\ 10^{-6}$	0	0.607
	LOC_Os12g25690.1|13112.m02604|protein	7	1.55	0.0006	1							1.55	0.0005977	1	
	Bradi1g19080.1	82	0.93	0.0290	0	0.70	0.0026	−1	0.94	0.2546	0	0.85	0.0010158	1	0.227
	Bradi2g56260.1	16	1.14	0.1313	0	1.16	0.0792	0	1.19	0.0764	0	1.16	0.011115	0	0.007
	Bradi1g29900.1	34	0.92	0.0501	0	0.91	0.0985	0	0.92	0.1725	0	0.92	0.0125685	0	0.004
	gi|291047856	11	1.20	0.0220	1							1.20	0.0219659	1	
	Bradi3g34980.1	14	1.28	0.0269	1	1.37	0.0277	1	0.98	0.7080	0	1.20	0.0285122	1	0.224
	Bradi5g01950.2	14	0.90	0.1622	0	0.88	0.1025	0	0.90	0.1496	0	0.89	0.0287141	0	0.003
	E2F3W4_WHEAT	4	1.32	0.0356	1							1.32	0.035621	1	
	B8YEL1_WHEAT	17	0.56			0.90	0.2091	0	0.82	0.0468	−1	0.74	0.0393589	1	0.175

C/Ref	Accession	#Peptides	Run 3 114/116	p-Value 114/116	Crit	Run 3 115/116	p-Value 115/116	Crit	Run 4116/117	p-Value 116/117	Crit	Geomean	Stouffers p-value	Trend value	t-Test p-value
Strong effect	Q5MCL9_WHEAT	12	1.63	0.0002	1	1.70	0.0000	1				1.66	3.38×10^{-8}	1	
	PSAA_WHEAT	17	0.21	0.0005	−1	0.30	0.0009	−1	0.31	0.0044	−1	0.27	9.092×10^{-8}	1	0.009
	Q45FE7_WHEAT	29	0.44	0.0003	−1	0.53	0.0007	−1	0.48	0.0087	−1	0.48	9.725×10^{-8}	1	0.005
	PSBA_WHEAT	41	0.29	0.0012	−1	0.48	0.0266	−1	0.40	0.0105	−1	0.38	1.315×10^{-5}	1	0.021
	Bradi5g16980.1	44	0.67	0.0014	−1	0.78	0.0306	−1	0.41	0.0130	−1	0.60	2.091×10^{-5}	1	0.120
	RUBA_WHEAT	56	0.64	0.0426	−1	0.64	0.0159	−1	0.68	0.0007	−1	0.65	2.324×10^{-5}	1	0.002
	A4UQP4_WHEAT	41	2.12	0.0057	1	1.39	0.0320	1	1.54	0.0078	1	1.66	4.312×10^{-5}	1	0.057
	Q5QJA5_WHEAT	13	1.91	0.0165	1	1.78	0.0028	1	1.75	0.0480	1	1.81	7.467×10^{-5}	1	0.002
	Bradi2g07940.2	3	1.97	0.0006	1	1.30	0.1010	0	2.02	0.0271	1	1.73	9.845×10^{-5}	1	0.063
	Q0IKM3_WHEAT	18	0.42	0.0063	−1	0.52	0.0269	−1	0.44	0.0237	−1	0.46	0.0001086	1	0.007

Each panel shows the top 10 protein in increasing order of the combined Stouffer p-value. The number of peptides is consolidated from the first run only. The first table shows a comparison where the effect is subtle: few differentially expressed proteins, lower ratios. The second table shows a comparison where the effect is strong: many differentially expressed proteins, large ratios, and good consistency. Care should be taken to include replication, for instance by requiring a minimum number of ratios per comparison. High consistency is not rewarded by the Stouffer approach but is rewarded by a one sample t-test, as apparent for instance in rows 4 and 5 of the first panel. More variability is tolerated by the Stouffer approach as the trend only is of interest, as apparent for instance in rows 5, 7 and 9 of the second panel

3. Decide on a cutoff that is acceptable for determining differential expression for the experiment at hand. Based on past experience a cutoff as low as 1.2 can be used if peptide level *p*-values are taken into consideration. In the presence of technical replicates, the false positive rate for the selected cutoff can be determined by identifying the number of proteins that pass the differential expression criterion when applied to technical replicates as a proportion of the whole data (*see* **Note 2**).
4. Using as criteria the *p*-value and ratio cutoff selected before, decide on the differential expression of each ratio. A protein ratio with ratio greater than the ratio cutoff (respectively, less than 1/cutoff) and peptide *p*-value less than the *p*-value cutoff is upregulated (respectively, downregulated); all other ratios are unmodified. The value gets recorded as −1, 0, or 1; see columns "Crit" in Table 1.
5. For each comparison, determine the trend value of the respective proteins, which records whether the combined ratios were in trend agreement or not. The values are in trend agreement if no opposing regulation trends are present (+1 for one ratio and −1 for one ratio), and at least one ratio is either up- or downregulated (*see* **Note 3**). Hence implicitly a trend value of 1 means that at least one ratio passes the ratio cutoff and is identified by at least two peptides.
6. Additionally, since it is applicable under the same circumstances, for each protein and each comparison, one can generate the one sample *t*-test *p*-value (Table 1 column *t*-Test). In most cases there should be a reasonable concordance. However we could see lower Stouffer *p*-values when the ratios are high, in same trend but more variable, as apparent in several rows in Table 1 panel 2. Thus the Stouffer approached is meant to reward a higher number of peptides, and not punish extra variability, assuming that the trend is more reproducible especially across different iTRAQ runs than the precise value.
7. For each comparison, decide differentially expressed proteins to be those having a Stouffer *p*-value less than cutoff (typically 0.05) and that are in trend (TrendVal = 1). Note that this guarantees that at least one ratio passed the ratio cutoff, and that at least 2 peptides are used for its identification, however does not necessarily as such guarantee replication, since a single ratio might have been combined.
8. One should be careful to note the number of ratios combined, as the conditions could be met for a single ratio only (Example Table 1, several proteins identified with one ratio only) (*see* **Note 4**).

3.5 Consistency Percentage

Calculating the consistency percentage is not necessary for determining differential expression, but gives a sense of the strength and consistency of the data for each comparison, and is similar to checking the correlation of the ratios.

1. For each comparison (e.g., A to Ref, or B to Ref, or C to D), determine the overall number of proteins identified as differentially expressed above, N_{DE} = (number of proteins with Stouffer combined p-value < 0.05 and TrendValue = 1) (*see* **Note 5**).
2. For each comparison, determine the overall number of proteins with Stouffer p-value less than cutoff N_S = (number of proteins with Stouffer combined p-value < 0.05); as above, can choose at least 2 or 3 replicates.
3. Calculate Consistency percentage = 100 × N_{DE}/N_S—*see* as an example Table 2 for the comparisons selected. In a comparison of two conditions C_1 and C_2 where there is little consistency in the ratios C_1/C_2, a consistency percentage of around 50% would be expected, as ratios could be randomly in trend or out of trend. In a highly consistent scenario, such as technical replicates, or similar biological replicates such as pools of plants, percentages over 80–90% could be obtained.
4. For comparison, calculate correlations for the combined ratios C_1/C_2; an example for a few selected comparisons is shown in Fig. 3. In general highly correlated ratios will yield a high consistency percentage (*see* **Note 6**). However, since the consistency percentage only includes ratios that have peptide level p-values generated, thus were identified with at least two peptides in one run, the consistency percentage can be quite high even when the correlation is not strong, as trend agreement rather than precise value agreement is rewarded.

3.6 Combining Multi-run Data in the Presence of a Redundant Database

A separate aspect to consider in the practical integration of iTRAQ data is how to handle protein grouping across multiple runs. The principle of parsimony [8] states that the number of reported proteins should not be overstated by double counting shared peptides; rather proteins identified by the same set of spectra should be grouped. This approach is taken by software such as Protein Pilot or Proteome Discoverer, which generates protein groups, with one or several proteins identified as winners. In a simplistic approach when working with a single iTRAQ run, it is acceptable to report the winners only, though for comparing with external literature the competitors may also need to be considered. In a multi-run scenario with a redundant database, the problem is that neither the protein group nor the protein winner is consistent across iTRAQ runs [16]. There are proposed solutions such as

Table 2
Example consistency percentage and correlation, with no restrictions (even 1 ratio is allowed on its own), or with three ratios required

		1 or more ratios			3 or more ratios		
Comparison	Correlations	Stouffer *p*-value < 0.05 and in trend	Stouffer *p*-value < 0.05	Consistency percentage (%)	Stouffer *p*-value < 0.05 and in trend	Stouffer *p*-value < 0.05	Consistency percentage (%)
A/Ref	−0.05 0.06 0.71	9	14	64.29	5	10	50.00
B/Ref	0.69 0.33 0.27	46	48	95.83	35	37	94.59
C/Ref	0.57 0.58 0.84	72	72	100.00	50	50	100.00
D/Ref	0.88 0.69 0.71	125	131	95.42	84	90	93.33

If only one ratio is used, the consistency may be overstated as a single ratio will always be in trend

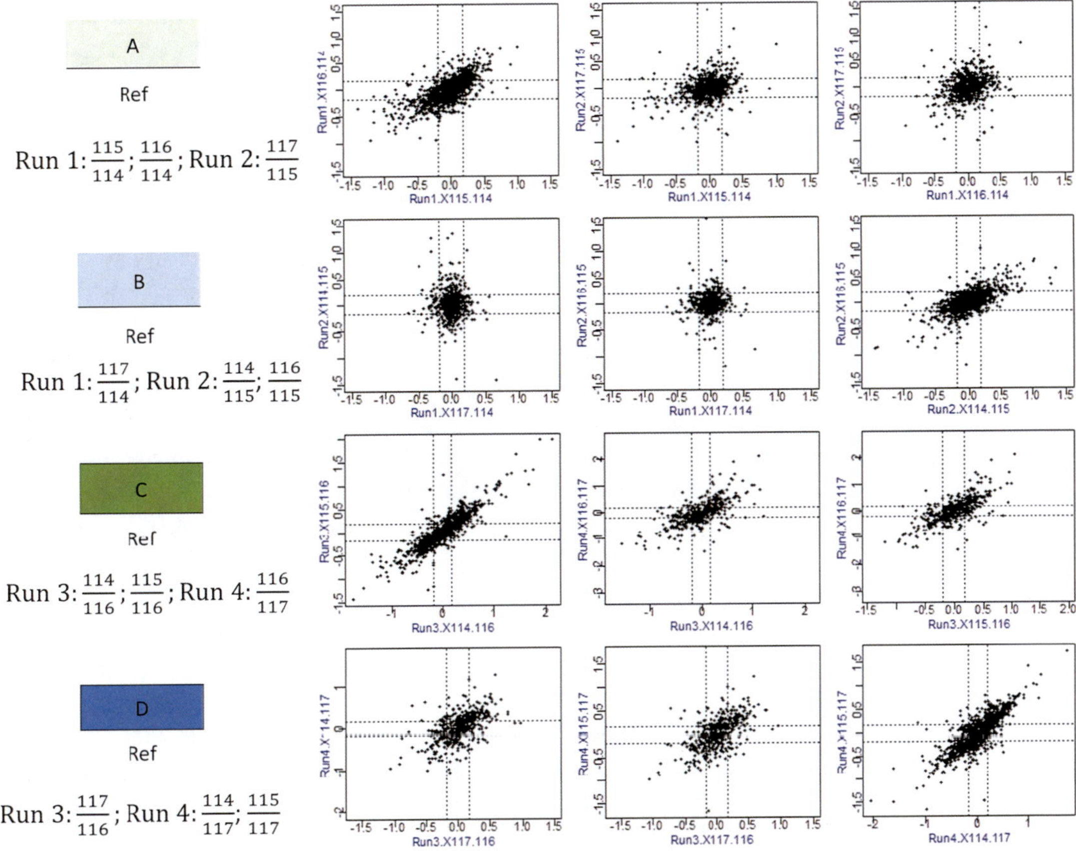

Fig. 3 Ratio correlation for some selected comparisons; note correlations across runs are typically poorer. Correlations are somewhat related to the consistency percentage, in that highly correlated ratios for a comparison (such as C/Ref or D/Ref) are likely to generate a high consistency percentage as shown in Table 2. However, since the consistency percentage is calculated using only ratios that pass the fold change criterion and the *p*-value threshold in some run at least, hence are identified by more than two peptides at least in one run, the consistency percentage can be quite high even if the correlations are only moderate, for example for comparison B/Ref

notably the ABSciex protein alignment template, or other approaches such as the Protein Group Code Algorithm [25] but they are not easily implemented; for instance the protein alignment template requires a master search as a reference frame which may be difficult to run for a large experiment. Once again, the easy solution of retaining the winners only (naïve overlap) may be too simplistic, and the more thorough solutions may be computationally challenging. Below we describe the fill-down option which is once again a pragmatic compromise, one that first evaluates if the naïve overlap is very different from the competitor or group-based overlap, then implements a fill-down option while still retaining a nonredundant set of proteins across the runs considered.

1. In order to see whether considering the competitors or not makes a big impact in a particular experiment we evaluate the run overlap of two separate iTRAQ runs with or without competitors. To calculate the naïve overlap, the competitors are simply discarded (*see* **Note 7**), and the overlap of the remaining winner proteins only can be easily calculated using any Venn diagram software.
2. To calculate the competitor overlap of two runs while still retaining only the minimum number of proteins across the two runs, first determine the set of proteins that are winners in at least one run. Then for each iTRAQ run, retain either the winner proteins, or any competitor proteins that have been selected as a winner in another run. Then determine the overlap of the two runs extended to include some competitors as described, again using any Venn diagram software.
3. If the naïve and competitor overlaps are quite similar, then neglecting the competitors has only a small impact in the experiment at hand. In the presence of a highly redundant or multi-species/multi-strain database, the competitor overlap is likely to be a lot larger than the naïve one. An example is show in [16] where the naïve overlap is 42% of the smaller run, but becomes nearly 70% when competitors are considered.
4. If the competitor overlap is considerably larger, then competitors can be retained, and a fill down procedure can be employed to assign the same quantitation to them as to the winner protein in the same run, though no significance value is assigned and the quantitation is marked as inferred via fill down. Figure 4 describes this procedure visually.
5. To fill down protein quantitation for a multi-run experiment, first determine the proteins that are group winners in at least one run; this will be the complete set of identifications across runs. For example in Fig. 4, the proteins retained as winners in some run are marked with a solid triangle. If no fill down is undertaken and the competitors are discarded, then in Fig. 4 the winner proteins would appear to be identified one in each run only. Now in each run, allow the competitors that are winners in some run (not necessarily the same run), and fill down the quantitation from the winner in each respective group. In the example of Fig. 4, the same parsimonious set of proteins is retained, but the quantitation is less sparse, and the proteins no longer appear to have been identified in one run only.
6. The filled down quantitation can be combined with the other computational approaches (ANOVA or pairwise comparisons), or can be left for manual examination/screening of individual candidates of interest.

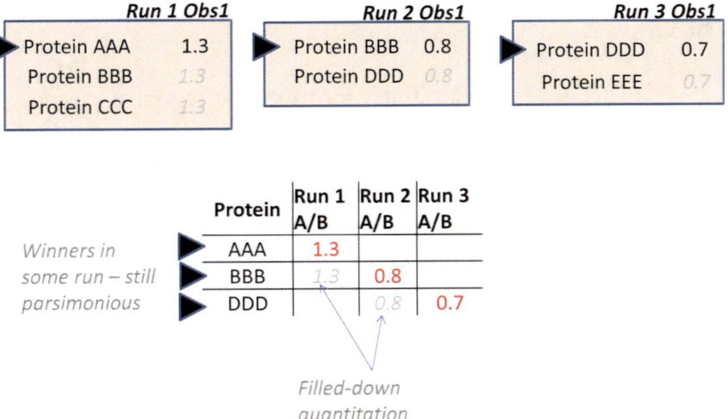

Fig. 4 Example of fill down for a possible set of protein groups; the proteins marked with black triangles are winners in some run, hence are all included in the combined set. Competitors such as CCC or EEE are not part of the final set. Quantitation is filled down for competitors BBB in Run 1 and DDD in Run 2, and marked in gray. The combined set with filled down quantitation is less sparse than the naïve overlap, but equally parsimonious in terms of identifications

3.7 Alternative Approaches

We have described here a relatively simple approach for determining differential expression in multi-run experiments, in situations where pairwise differential expression is of interest. Such an approach can also be used in the context of a single run, if the interest is in emphasizing more abundant proteins whose quantitation may be more reliable. The main advantage of such an approach is its computational simplicity, the fact that it takes protein quality into consideration, and will lead to more abundant/reliable protein identified as differentially expressed. Likewise, it will not punish minor variations in quantitation across runs, but emphasize instead trend agreement.

Among its drawbacks are the fact that it cannot handle all possible experimental designs, and will only extract protein ratios when the samples compared are placed on the same run—hence in some situations it may be deemed as less complete, though in the presence of a common reference it can be used in combination with a multivariate approach on the consolidated protein ratios. In a complex experiment there are almost always multiple avenues of analysis, and to reflect that a multitude of tools have been developed for the analysis of isobaric labeled data [18, 26–28].

Additionally, we have touched briefly on the problems presented by redundant databases, and described possible pragmatic approach to first consider whether database redundancy will have a large influence on the experiment at hand, and if so gave a simple strategy for reducing the resulting potentially poor run overlap.

4 Notes

1. The geometric mean, also known as "mean in log space," ensures that the when considering the mean of B/A we will obtain 1/(mean(A/B)), as expected. The arithmetic mean would not have this property.
2. An example of an FDR calculation based on replicates is included in the supplementary information of [21].
3. Note that two ratios could have opposing trends as long as they don't also have low *p*-values (less than the *p*-value cutoff). Thus in trend ratios cannot have opposing trends both with high confidence.
4. Individual analyses may have different appropriate levels of replication. Though balanced designs are often employed, in some cases some comparisons might be of lesser interest than others, and could have less replication in an iTRAQ experiment. The lowest number of ratios acceptable should be recorded.
5. As in the previous note, the acceptable level of replication should be decided as appropriate for each project. Clearly the requirement of being "in trend" is always satisfied if a single ratio is being combined.
6. In general, comparisons of two conditions where the effects are subtle (low ratios, few differentially expressed proteins) will yield low correlations.
7. In ProteinPilot exports, the competitors can be easily filtered by the Unused score: competitor proteins are assigned an Unused score of 0.

References

1. Ross PL, Huang YN, Marchese JN et al (2004) Multiplexed protein quantitation in *Saccharomyces cerevisiae* using amine-reactive isobaric tagging reagents. Mol Cell Proteomics 3:1154–1169
2. Thompson A, Schäfer J, Kuhn K et al (2003) Tandem mass tags: a novel quantification strategy for comparative analysis of complex protein mixtures by MS/MS. Anal Chem 75:1895–1904
3. Gillet LC, Navarro P, Tate S et al (2012) Targeted data extraction of the MS/MS spectra generated by data-independent acquisition: a new concept for consistent and accurate proteome analysis. Mol Cell Proteomics 11:O111.016717
4. Braun CR, Bird GH, Wühr M et al (2015) Generation of multiple reporter ions from a single isobaric reagent increases multiplexing capacity for quantitative proteomics. Anal Chem 87:9855–9863
5. Boehm AM, Pütz S, Altenhöfer D et al (2007) Precise protein quantification based on peptide quantification using iTRAQ. BMC Bioinformatics 8:214
6. Shadforth IP, Dunkley TP, Lilley KS et al (2005) i-Tracker: for quantitative proteomics using iTRAQ. BMC Genomics 6:145
7. Arntzen MØ, Koehler CJ, Barsnes H et al (2010) IsobariQ: software for isobaric quantitative proteomics using IPTL, iTRAQ, and TMT. J Proteome Res 10:913–920

8. Nesvizhskii AI, Aebersold R (2005) Interpretation of shotgun proteomic data: the protein inference problem. Mol Cell Proteomics 4:1419–1440
9. Rauniyar N, Yates JR 3rd (2014) Isobaric labeling-based relative quantification in shotgun proteomics. J Proteome Res 13:5293–5309
10. Ow SY, Salim M, Noirel J et al (2009) iTRAQ underestimation in simple and complex mixtures: "the good, the bad and the ugly". J Proteome Res 8:5347–5355
11. Karp NA, Huber W, Sadowski PG et al (2010) Addressing accuracy and precision issues in iTRAQ quantitation. Mol Cell Proteomics 9:1885–1897
12. Bräutigam A, Shrestha RP, Whitten D et al (2008) Low-coverage massively parallel pyrosequencing of cDNAs enables proteomics in non-model species: comparison of a species-specific database generated by pyrosequencing with databases from related species for proteome analysis of pea chloroplast envelopes. J Biotechnol 136:44–53
13. Kamath KS, Pascovici D, Penesyan A et al (2016) *Pseudomonas aeruginosa* cell membrane protein expression from phenotypically diverse cystic fibrosis isolates demonstrates host-specific adaptations. J Proteome Res 15:2152–2163
14. Padliya ND, Garrett WM, Campbell KB et al (2007) Tandem mass spectrometry for the detection of plant pathogenic fungi and the effects of database composition on protein inferences. Proteomics 7:3932–3942
15. Seymour SL (2010) Assessing and interpreting protein identifications. J Biomol Tech 21:S12
16. Pascovici D, Gardiner DM, Song X et al (2013) Coverage and consistency: bioinformatics aspects of the analysis of multirun iTRAQ experiments with wheat leaves. J Proteome Res 12:4870–4881
17. Hill EG, Schwacke JH, Comte-Walters S et al (2008) A statistical model for iTRAQ data analysis. J Proteome Res 7:3091–3101
18. Oberg AL, Mahoney DW, Eckel-Passow JE et al (2008) Statistical analysis of relative labeled mass spectrometry data from complex samples using ANOVA. J Proteome Res 7:225–233
19. Whitlock MC (2005) Combining probability from independent tests: the weighted Z-method is superior to Fisher's approach. J Evol Biol 18:1368–1373
20. Pascovici D, Song X, Solomon PS et al (2015) Combining protein ratio *p*-values as a pragmatic approach to the analysis of multirun iTRAQ experiments. J Proteome Res 14:738–7461
21. Winterberg B, Fall LAD, Song X et al (2014) The necrotrophic effector protein SnTox3 reprograms metabolism and elicits a strong defence response in susceptible wheat leaves. BMC Plant Biol 14:215
22. Ullrich M, Liang V, Chew YL et al (2014) Bioorthogonal labeling as a tool to visualize and identify newly synthesized proteins in *Caenorhabditis elegans*. Nat Protoc 9:2237–2255
23. Song X, Bandow J, Sherman J et al (2008) iTRAQ experimental design for plasma biomarker discovery. J Proteome Res 7:2952–2958
24. Martinez-Val A, Garcia F, Ximénez-Embún P et al (2016) On the statistical significance of compressed ratios in isobaric labeling: a cross-platform comparison. J Proteome Res 15:3029–3038
25. Freue GVC, Sasaki M, Meredith A et al (2010) Proteomic signatures in plasma during early acute renal allograft rejection. Mol Cell Proteomics 9:1954–1967
26. Navarro P, Trevisan-Herraz M, Bonzon-Kulichenko E et al (2014) General statistical framework for quantitative proteomics by stable isotope labeling. J Proteome Res 13:1234–1247
27. Choi M, Chang CY, Clough T et al (2014) MSstats: an R package for statistical analysis of quantitative mass spectrometry-based proteomic experiments. Bioinformatics 30:2524–2526
28. Zhou C, Walker MJ, Williamson AJK et al (2013) A hierarchical statistical modeling approach to analyze proteomic isobaric tag for relative and absolute quantitation data. Bioinformatics 30:549–558

Chapter 14

Quantitative Proteomics Data in the Public Domain: Challenges and Opportunities

Andrew F. Jarnuczak, Tobias Ternent, and Juan Antonio Vizcaíno

Abstract

Mass spectrometry based proteomics is no longer only a qualitative discipline, and can be successfully employed to obtain a truly multidimensional view of the proteome. In particular, systematic protein expression profiling is now a routine part of many studies in the field and beyond. The large growth in the number of quantitative studies is accompanied by a trend to share publicly the associated analysis results and the underlying raw data. This trend, established and strongly supported by public repositories such as the PRIDE database at the European Bioinformatics Institute, opens up enormous possibilities to explore the data beyond the original publications, for instance by reusing, reanalyzing, and performing different flavors of meta analysis studies. To help researchers and scientists realize about this potential, here we describe the mainstream public proteomics resources containing quantitative proteomics data, including the processed analysis results and/or the underlying raw data. We then present and discuss the most important points to consider when attempting to (re)use proteomics data in the public domain. We conclude by highlighting potential pitfalls of (re)using quantitative data and discuss some of our own experiences in this context.

Key words Mass spectrometry, Data (re)analysis, Quantitative proteomics, Data repository, PRIDE database

Abbreviations

CPTAC	Clinical Proteomic Tumor Analysis Consortium
DDA	Data-dependent acquisition
DIA	Data-independent acquisition
iTRAQ	Isobaric tags for relative and absolute quantification
MIAME	Minimum information about a microarray experiment
MIAPE	Minimum information about a proteomics experiment
MRM	Multiple reaction monitoring
MS	Mass spectrometry
PRM	Parallel reaction monitoring
PSM	Peptide spectrum match

Caroline A. Evans et al. (eds.), *Mass Spectrometry of Proteins: Methods and Protocols*, Methods in Molecular Biology, vol. 1977, https://doi.org/10.1007/978-1-4939-9232-4_14,

PTM	Posttranslational modification
PX	ProteomeXchange
SILAC	Stable isotope labeling by amino acids in cell culture
TMT	Tandem mass tag

1 Introduction

The aim of proteomics is to describe the complete set of proteins in any given biological system, connecting the genotype and the phenotype in very different scenarios: for example, in a primary breast tumor sample obtained from a patient or in an industrially relevant strain of *Saccharomyces cerevisiae* growing in a standard culture medium. While obtaining a catalogue of the proteins present is a necessary first step, in order to fully understand the phenotype, it is important to establish not only the presence/absence of a given protein but also its abundance, its subcellular localization, the presence of posttranslational modifications (PTMs), and/or the interaction networks it might be involved in. In an ideal case, measuring several of those parameters simultaneously will allow for a truly multidimensional proteome analysis [1]. However, this is not always feasible in practice, and researchers usually focus on a small number of properties alone. In particular, determining protein abundance has become an indispensable part of the majority of state-of-the-art proteomics studies as well as in a multitude of functional and systems biology approaches.

Just to highlight a couple of recent examples, a study by Wang et al. measured protein expression patterns of human colorectal cancer samples, such as primary colorectal cancer cells and normal tissue, and showed that cell lines, often used to model the disease, represent primary tumors relatively well. The authors were also able to assess the utility of proteomics data to predict response to anticancer drugs, and demonstrated that it can improve the already available models [2]. In another example, this time from the model organism *Saccharomyces cerevisiae*, absolute protein quantification values of thousands of proteins and their corresponding transcripts were used to gain better insights into translational efficiency and the relationship between transcriptome and proteome [3, 4].

The explosion of quantitative proteomics studies can in part be attributed to substantial improvements in experimental methods for sample preparation [5, 6], data analysis [7, 8], and to the ground-breaking developments in MS instrumentation over the last decade [9, 10]. Closely following these developments, there has been a shift in the attitude in the proteomics community toward making the data associated with scientific studies more open and publicly available. This trend, already common in more mature disciplines such as genomics and transcriptomics, was largely triggered by the establishment of public repositories for

MS-based proteomics data, notably the PRIDE database [11, 12], and subsequent development of the ProteomeXchange (PX) Consortium of public proteomics resources [13, 14]. These initiatives provide a reliable infrastructure and facilitate easy storage, sharing, and reuse of public MS proteomics data. More than a decade later, the trend toward open science in proteomics does not show any signs of slowdown. In fact, more repositories are joining the ProteomeXchange Consortium [15], and major life sciences journals now mandate or strongly recommend data deposition as a condition of publication.

In addition to large archiving projects, which capture almost any aspect of the proteomics data (raw data, peptide and protein identification, quantification, and metadata), other related resources devoted to the reanalysis and aggregation of multiple studies are also available to the scientific community. In fact, the public proteomics data can be reused by others in many different ways, e.g., to combine multiple studies in order to achieve a single improved result (meta-analysis approaches), or to integrate proteomics data with data coming from other omics approaches [16, 17].

In this book chapter, we will discuss the main public resources that capture quantitative proteomics studies and their associated metadata, and explain how the data can be accessed and reused. We will also highlight some caveats concerning the reuse of the data.

2 Public Proteomics Resources Containing Quantitative Data

A wide range of public proteomics resources are currently available. A comprehensive review of the most relevant resources is available in [18]. Many of those focus exclusively on the peptide and protein identification level. For example, the Global Proteome Machine Database (GPMDB) stores the results of MS/MS spectra reanalysis including the associated peptide and protein identifications as well as PTMs or amino acid variants [19]. Another relevant example is the PX Consortium member PeptideAtlas project and its various components [20, 21]. Here, we turn our attention only to those public resources that provide access to quantitative proteomics data. Those can be further divided into two types of databases: (i) archival resources that store the raw data and analysis results, as provided by the original authors and (ii) secondary resources that reuse the data available in archival resources via their own analysis pipelines, providing their own interpretation.

2.1 Archival Resources

2.1.1 PRIDE Database

The PRoteomics IDEntifications (PRIDE) database (https://www.ebi.ac.uk/pride/archive/) [11, 12] is the largest resource dedicated to the storage and dissemination of MS proteomics data, and is a founder member of the PX Consortium. PRIDE contains raw experimental data and analysis (processed) results which are directly submitted by scientists from all over the world. The submitted results include peptide and protein identifications and, where relevant, quantification values. Processed peak lists, protein sequence databases or spectral libraries, and analysis scripts can also be deposited. Although the submissions are curated by PRIDE in order to ensure data integrity, they ultimately are made available as originally provided and annotated by authors. However, in cases where the dataset constitutes a reanalysis of other previously submitted public dataset, a "Reanalysis" tag is assigned to the dataset.

To facilitate data reuse, each project in PRIDE includes a description of the sample and data processing protocols. Software, quantification method, and many other metadata can also be made available, if provided by the submitters. In any case, the original publication (containing the full details) is linked to each dataset, which is very useful for those who want to reuse the data.

Importantly, PRIDE supports and encourages data submissions in standard open file formats developed by the Proteomics Standards Initiative (PSI), which highly improves accessibility and reusability of the data in the resource. Datasets containing these files represent the so-called "Complete" submissions, where it is possible for PRIDE to link the processed results (at least at the identification level) and the mass spectra. The two standard PSI file formats supported for the analysis results are mzIdentML [22] and mzTab [23]. In addition, mzML [24] and other peak list formats are supported for encoding MS data coming from the mass spectrometers. While mzIdentML is an XML-based format limited to identification results only, mzTab files can accommodate protein and peptide identifications as well as quantitative information and experimental metadata. The stand-alone tool PRIDE Inspector [25] can be used to visualize proteomics data available in all these open formats.

If processed results are not provided in these PSI standard formats, the datasets are called "Partial." In that case, all the files are still made available to download, although it is not possible to link identifications and mass spectra. There are two ways of accessing protein abundance values in PRIDE, either by downloading the results submitted by users in different formats and parsing/visualizing them, or by obtaining the raw data and reprocessing it. All the relevant files are easily available through the PRIDE FTP site.

2.1.2 Mass Spectrometry Interactive Virtual Environment (MassIVE)

MassIVE (http://massive.ucsd.edu/) is a free resource developed in UCSC (University of California, San Diego) to facilitate submission and dissemination of MS data. Also a member of the PX Consortium, much alike PRIDE, MassIVE accepts raw experimental data as well as processed analysis results and additional metadata. Same as in PRIDE, "Complete" and "Partial" datasets also exist.

Additionally, MassIVE makes available a number of additional resources, for example, online analysis workflows to reanalyze publicly available MS/MS spectra, or to compare identification results between datasets. It is also possible to attach the reanalysis results or to add comments on datasets, providing a "live" version of all existing datasets. As in case of the PRIDE, access to quantitative information can be obtained by downloading the analysis results as provided by data submitters, or by directly reanalyzing the raw files associated with a given project.

2.1.3 The CPTAC Data Portal

The Clinical Proteomic Tumor Analysis Consortium (CPTAC) is an effort to enhance the understanding of the molecular mechanisms of cancer using proteomic technologies and workflows in combination with genomics and transcriptomics information [26]. The Consortium produces large proteomics datasets from cancer biospecimens, human-in-mouse cancer tumor xenografts or clinical tumor and normal samples, many of which had been previously characterized through an application of genomics technologies. The work of the Consortium has already resulted in a number of integrated proteogenomics studies [27, 28]. Importantly, data analysis in the project is performed using the CPTAC Common Data Analysis Platform [29]. This pipeline produces both peptide-spectrum-match (PSM) centric, and gene centric reports, minimizing the variability, due to data processing differences that might otherwise arise from various analysis sites.

The CPTAC data portal (https://proteomics.cancer.gov/data-portal) was created as a centralized repository for the data generated by the Consortium [30]. The portal allows free access to the raw MS files as well as to the analysis results (including the mapping of spectra to peptide sequences and protein identifications, and technical and clinical metadata).

Across the many CPTAC studies, relative protein quantification was performed using a variety of methods, including spectral counting, label-free or reporter-based ion quantification (isobaric tags for relative and absolute quantification, iTRAQ). Quantification reports are provided in a simple tab-delimited format.

2.2 Secondary Resources

2.2.1 ProteomicsDB

ProteomicsDB [31] (https://www.proteomicsdb.org/) is a publicly available and actively developed database dedicated to the exploration and quantitative analysis of human proteomics data. It is the primary resource that was developed to accommodate one of the two drafts of the human proteome [31]. Initially, ProteomicsDB consisted of a raw data repository and a set of visualization and analytical tools. However, as of January 2017, the repository part was discontinued and, at the moment of writing, all data will be soon removed from there, since most of it is already available in PRIDE. Another recent development is the integration of almost three million reference spectra from the ProteomeTools project

[32]. ProteomeTools is a recently published large-scale effort to synthesize and analyze peptides mapping to human genes using LC-MS/MS and various peptide fragmentation methods.

In ProtemicsDB, raw data is reanalyzed using the popular software tools MaxQuant [8] and Mascot [33] (http://www.matrixscience.com/). At the moment of writing there is already coverage for 15,721 human proteins and 113,944 unique peptide isoforms there. The results can be accessed through a rich web interface by browsing for peptide sequences or protein names in the "Human Proteins" and "Peptides" tabs, respectively. The "Chromosomes" tab shows the protein coverage by individual chromosomes.

Protein abundance values can be accessed through the "Expression" tab in the "Human Proteins" view and are annotated for different tissues, cell lines, and body fluids. A map of the human body can visualize the abundance levels across the different tissues. In addition, a histogram displayed on the page shows the median protein expression and corresponding standard deviation, where replicate information is available. Users have the choice of selecting different quantification methods: MS1-based (iBAQ and Top3) or MS2-based. Protein ratios can be calculated based either on iBAQ or Top3 intensities. mRNA expression data, downloaded from the Human Protein Atlas [34], is also currently integrated with proteomics data.

The "Analytics" window offers alternative expression analysis in the form of a heat map, which is constructed for a set of protein entries. Then data are clustered based on the corresponding expression values. A choice of quantification methods is also available in this view.

2.2.2 Human Proteome Map

The "Human Proteome Map" (http://humanproteomemap.org/) is a web-based resource accompanying the second draft of the human proteome [35]. While the web portal allows the visualization and exploration of protein abundance levels in human fetal and adult tissues and hematopoietic cells as reported in the original paper, the MS data underlying the results is deposited in PRIDE (accession number PXD000561).

The results available in the Human Protein Map are static and were obtained by searching the MS/MS spectra against NCBI's Human RefSeq database using the Sequest [36] and Mascot [33] search engines. Quantification was then achieved by calculating a normalized spectral counting index, i.e., the number of spectra for each gene in each experiment was summed across all peptides mapping to that gene and normalized by the total number of acquired spectra. Protein abundance values can be visualized as heat maps in this resource. Those can be plotted for a user-defined gene list or for all proteins present in the provided predefined lists of pathways or gene families. A choice to base the quantification on shared and/or unique peptides or in only gene-specific peptides is also available. The resulting plots can be downloaded, and the full set of results, including the gene, protein, and peptide-level expression matrices, are available as Excel spreadsheets or as comma-separated text files.

2.2.3 PaxDb

PaxDb (http://pax-db.org/) is a resource providing absolute protein expression levels across different organisms and tissues [37]. PaxDb aggregates information coming from independent proteomics studies, including biochemical, biophysical, and label-free MS quantification experiments. The data is obtained directly from the published literature, and from public repositories such as PRIDE or PeptideAtlas.

Importantly, the data is curated and reanalyzed in order to make comparable the results of various studies. Protein identifications are mapped into a common identifier space based on the reference genomes, as obtained from the STRING database [38]. Protein abundances are then calculated in "parts per million" (ppm), i.e., by expressing the protein abundance as a proportion of all protein molecules available in a given dataset, and re-scaled to 1,000,000. Overall, ppm scaling has the advantage of providing more consistent values across different techniques and provides efficient normalization across different datasets. Furthermore, datasets are also scored to estimate their quality. As a meta-resource, PaxDB allows access to the results from individual studies as well as from aggregated datasets. Protein abundance values can be downloaded in tab-delimited files. At the time of writing, the latest PaxDb release (year 2015) made use of 414 datasets coming from 53 organisms, and covered abundance values for more than 300,000 proteins [39].

2.2.4 MaxQB

MaxQB (http://maxqb.biochem.mpg.de/mxdb/) is a resource that allows access to information including the joint analysis and comparison of large-scale proteomics projects (currently 17 projects) coming from the Max Planck Institute of Biochemistry [40]. The underlying raw data was acquired on high-resolution Orbitrap instruments and was consistently analyzed using the MaxQuant computational proteomics platform [41]. While most of the corresponding raw data has been made available in the PRIDE database, MaxQB stores the detailed output of the analyses, including peptide and protein identifications, the corresponding tandem mass spectra, and protein abundance information. Additional metadata can also be stored, such as the experiment name and description, the sample source, perturbation, the organism, or labeling method. Detailed MaxQuant processing parameters and a processing report are also made available. Protein quantification values are provided in different ways depending on the project setup and labeling method: e.g., as log10 Intensities, log10 iBAQ values, relative expression expressed as parts per million, or SILAC ratios. All data is downloadable in a tab-delimited format.

3 General Considerations of Quantitative Proteomics Experiments

Proteomics is a relatively young discipline and new experimental approaches, instrumentation, acquisition methods, and analysis software are continuously developed. As no single procedure is applicable to all cases, scientists have a broad choice of up-to-date techniques they can employ in a given scenario. However, there is usually no single "gold standard" procedure which could be applied in that particular scenario, unlike some other more mature omics fields where a broader consensus is accepted for what constitutes best practice [42]. Also, quality control (QC) procedures are not well defined yet in terms of experimental protocols, measurements of technical variability or about what metrics to feed back to users. However, notable steps have been made to address these problems [43], such as the recent establishment of the HUPO PSI QC Working Group [44] (http://www.psidev.info/groups/quality-control).

A wide variety of protocols exists, many with only minor alterations. This is particularly evident in quantitative proteomics and the related downstream data processing and analysis, including the statistical analysis. Quantitative proteomics experiments can vary greatly in:

1. *Quantification method.* Some of the most common methods include: label-free approaches including spectral counting, where the number of MS/MS spectra is used to provide a quantitative read out, and extracted ion chromatogram (XIC)-based quantification where an integrated area of a peptide peak is obtained. In addition, isotopic-labeling strategies (in particular stable isotope labeling by amino acids in cell culture (SILAC) and iTRAQ), which are based on a principle of comparing abundance values between two or more differentially labeled samples. Comprehensive reviews of the many quantification techniques are available elsewhere [45, 46].
2. *Instrumentation and data acquisition modes.* Proteomics is driven by MS technology. Many instrument types are currently employed by researchers, including quadrupole time-of-flight, linear ion trap, Orbitrap and ion mobility devices, as well as various hybrids thereof. Each offers varying levels of sensitivity, resolution, mass accuracy or dynamic range, which will inevitably affect quantification results—making direct comparisons difficult. In addition, different acquisition modes can be employed: data-dependent acquisition (DDA), multiple reac-

tion monitoring (MRM), parallel reaction monitoring (PRM), or data-independent acquisition (DIA), to name just the most popular ones [47–49].

3. *The number of identified and quantified proteins.* The percentage of peptides successfully identified (and therefore later quantified) in a proteomics experiment greatly depends on the software tools employed and their settings. Here, the database search algorithms play a crucial role. In general, peptide and protein identification are achieved by search engines, which aim to match peptide spectra from in silico digests with those measured experimentally, search a spectral library (comparing experimental spectra with collection of previously generated spectra) or derive the sequence based on the MS/MS spectrum alone (de novo approaches). While most search engines appear to perform satisfactory on their own, each algorithm can produce unique sets of peptide and protein identifications. Although, direct comparisons of the performance of search engines are difficult, some have been attempted, and those reinforce the importance of parameter tuning and careful parametrization [50, 51]. Additionally, even searches performed with the same search engine can result in different rates of identifications. For example, a large dataset searched against a custom curated human proteome with selected sequences (e.g., UniProtKB/SwissProt) will yield a different number of hits to a search against a much more expanded, non-reviewed and automatically annotated database like UniProtKB/TrEMBL. This point is illustrated in Fig. 1, where an overlap between the number of identified peptides and successfully quantified protein groups is displayed for two searches using the same data. Search 1 was performed using UniProtKB/SwissProt (a highly curated canonical human proteome database containing 21,044 sequences), and search 2 using the UniProt reference human proteome (70,958 sequences).
4. *Final quantification results.* Similarly to identification, quantification results can be massively affected by the use of different processing pipelines and normalization methods. Here, comprehensive and a "like for like" comparisons are also difficult. Normalization, in particular, is an important step in proteomics data analysis where the aim is to accurately describe the biological effects under investigation. This is because experiments, no matter how well controlled and executed, will always have some level of inter-sample variability and technical noise. The sources of this bias can be multiple: from temperature fluctuations, small differences in sample processing like pipetting errors or differences in reagents quality, to variability of spray stability and liquid chromatography conditions. Normalization is the process of removing these biases while retaining the bio-

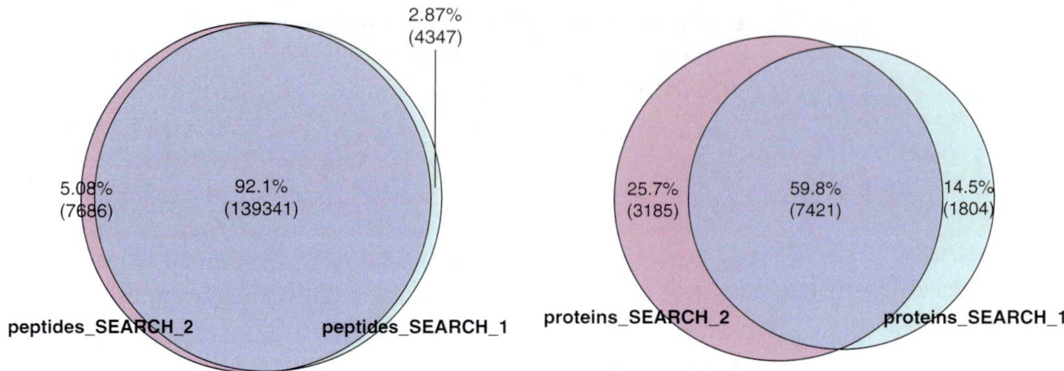

Fig. 1 Identification greatly depends on the database being searched. Overlap between the identified peptide and protein groups searches using two recent protein sequences databases: (SEARCH_1) UniProtKB/SwissProt, human canonical proteome (21,044 sequences) and (SEARCH_2) UniProt reference human proteome (70,958 sequences). RAW files from dataset PXD000612 were downloaded from the PRIDE database (http://www.ebi.ac.uk/pride/archive/projects/PXD000612) and analyzed using MaxQuant with default search parameters

logically relevant signal and making samples comparable. Several normalization methods are currently popular and widely used in the field, and each will have a different effect on the analysis results [52–55]. Furthermore, as the variety of normalization methods can be overwhelming, some good tools to evaluate and select the best performing approaches are also available [56, 57].

5. *File formats.* The diversity and complexity of proteomics instrumentation and analysis workflows is closely followed by the large diversity in file formats used to store raw data and analysis results. Mass spectrometers from different manufacturers and vendors save data to closed proprietary binary formats, and nearly all of the software packages used for data analysis (open source or analysis ecosystems provided by the vendors) write output in a different file format. This can seriously hamper large-scale reanalysis efforts, because data is difficult to manage and manipulate. Simple tasks like retrieving MS/MS scans from various raw files, or lists of peptide spectrum matches from multiple results files can become laborious and time consuming. Additionally, the access to the binary vendor formats is at present by large confined to the Microsoft Windows^R operating system, as the libraries provided by the vendors make it a requirement. Some exceptions are starting to emerge, for example, from version 1.6.0.1, MaxQuant [8] supports cross platform data processing on Windows, Ubuntu, and the Mac OS operating systems. It is also important to highlight that significant efforts have been extended to make raw data available in open data formats, such as mzXML, developed at the Institute of Systems Biology [58], and more

recently the PSI standard mzML [24]. In addition, ProteoWizard library and tools, including the MSConvert GUI tool, emerged as reliable and widely used toolkit to convert the binary formats to variety of open formats [59, 60].

Other file formats, for performing downstream data processing calculations, storing analysis results or to facilitate data visualization have also been implemented, and are often accompanied by open-source software libraries [25, 61]. Of those, the most popular formats containing data and metadata generated by the identification and quantification analysis software are mzIdentML [22] (mentioned above, for identification data only), mzQuantML [62] (for quantification information), and mzTab [23] (also mentioned above, for both identification and quantification).

4 Pitfalls of (Re)Using Quantitative Proteomics Data

A large number of public quantitative proteomics datasets are currently available for researchers. However, in order to make an efficient use of this data one must consider the limitations outlined in the previous section. Additionally, many of the considerations will depend on the specific aims of the study. For example, different use cases such as the direct comparison of results, validation of experimental or computational methods, the combination of multiple studies to achieve improved results (meta-analysis studies), or the integration of proteomics data with other omics data types will each require a different approach and assumptions in the interpretation of the results. Nonetheless, many of the considerations will be universally applicable, and here we highlight those.

4.1 Direct Comparison of the Reported Results: Single Study (Re)Use Versus Multiple Studies

The first use case is the reuse of quantitative results directly as reported by the original studies. The initial consideration is whether protein/peptide expression is going to originate from a single study or multiple studies. By a single study we mean that the samples under investigation were prepared using the same experimental technique in the same experimental context, data was generated on a single instrument and processed with one analysis software. In contrast, multiple studies could involve the same sample but different experimental parameters thereafter. Because proteomics MS experiments are not inherently quantitative, i.e., there is no simple linear relationship between concentration or amount of peptides and signal response of the instrument [63] and quantitative intra- and inter-laboratory reproducibility can vary considerably [64], even what in principle would appear to be highly similar studies are not always directly comparable.

To help illustrate this point we have obtained relative quantification values from two studies performed in *S. cerevisiae*, as reported by the authors in [65, 66]. Both studies measured the proteome response of yeast to osmotic shock by addition of NaCl (0.4 and 0.7 M, respectively) during a time course. In both studies *S. cerevisiae* strain BY4741 was used, and samples were removed at four common time points: 30, 60, 90, and 120 min after the addition of salt. The protocols followed were different though: after cell lysis, protein digestion, different quantification approaches (Tandem Mass Tag (TMT) labeling vs. label-free approach), data acquisition in different instruments (LTQ Orbitrap Velos vs. 5600 TripleTOF) and computational analysis, fold changes were calculated between the time points 0/30 min, 0/60 min, 0/90 min and 0/120 min. Both studies were performed in biological replicates to minimize the effects of outliers or technical noise, and are effectively internally standardized to the time point 0 min.

In theory, the results of the two studies should be highly comparable. However, when plotting the correlation of the reported fold changes in both studies (Fig. 2) it can be seen that this is not always the case. The correlation between individual time points within one study is high (Pearson correlation, $R > 0.8$), suggesting a high reproducibility. An inter-study comparison, however, reveals a much higher degree of divergence ($R < 0.4$) in the quantification results. This striking disparity most likely arises from multiple factors: differences in the sample preparation, data acquisition, and data processing workflows.

It is therefore important to exercise caution when attempting to reuse quantitative results directly from multiple studies. The usual alternative approach is to perform a full reanalysis of the raw data using the same analysis pipeline, including a systematic normalization to account for inter-study biases, or to combine only the results of the statistical analyses (e.g., *p*-values) rather than the abundance/fold change values directly reported by the different studies. This is the approach followed for example in [31] (ProteomicsDB, see previous section), where the raw data was reanalyzed using MaxQuant, and multiple label-free quantification methods were benchmarked against a "ground truth" set of protein copy-number estimates, in order to determine the best performing method.

4.2 Incomplete or Erroneous Data Annotation

Once the relevant studies and the corresponding publications are identified, the underlying data should be sufficiently described in order to facilitate a meaningful reuse. The notion of technical and biological metadata is to provide such systematic descriptions. Unfortunately, the lack of appropriate metadata in proteomics repositories is a significant hurdle all too often encountered in omics meta- or reanalysis studies [67, 68]. One of the main reasons for this is that the software used to generate proteomics results is

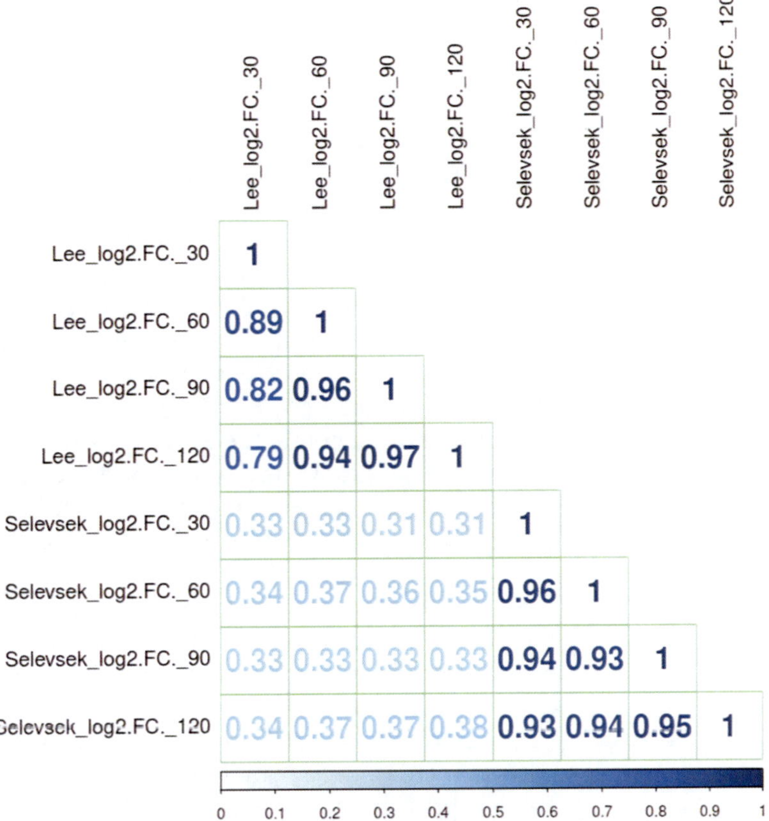

Fig. 2 Correlation within and across studies. Correlation matrix of the reported fold changes coming from two studies of osmotic shock in yeast [65, 66]. Each cell in the plot shows the Pearson correlation values between two measurements (time points after osmotic stress). For example, the bottom left cell, $R = 0.34$, represents the correlation between fold changes reported in the Lee et al. study at 30 min (Lee_log2.FC_30) and those reported in Selevsek et al. at 120 min (Selevsek_log2.FC_120). The darker the blue color, the higher the correlation found. It can be seen that the correlation between fold changes within the same study is much higher than between different studies. The comparisons are made based on fold changes of 1233 proteins quantified in both studies, and in all time points

by large unaware of the metadata associated with the analyzed sample. Thereby, even if the results are submitted using a standard file format, the initially generated files contain very little metadata (if any) about the sample and/or the experimental design.

If this information is not present in the submitted files, it needs to be provided by the submitters. In our experience, there is always a balance between the required amount of metadata and the willingness of researchers to submit their data. This balance was taken into account when designing the current metadata requirements for the members of the PX Consortium in 2011–2012. The focus was put on making it as practical and easy as possible for researchers to make their data publicly available and accessible. This was clearly needed at the time, since the primary objective was to change the "culture" of data sharing in the field and public data deposition was still scarce. In this context, annotating processed result files is seen as additional work for the submitter—work that, in most cases, is not perceived to be of direct benefit to them. As a result, only a part of the metadata from the analyzed datasets is available in an organized, machine-readable structure.

The problem can vary in severity. The lack of any metadata obviously excludes a study from almost any form of reuse, while metadata that is incomplete might complicate or limit the interpretation of results. In other cases, incorrect annotation will likely result in erroneous or invalid conclusions. The only solution at present is to curate the data manually, by reading in detail the original publication (including the supplementary material), or even contacting directly the authors. In the near future, metadata requirements will become stricter. This will need to happen gradually, due to the resistance of submitters, but also due to the need to extend current submission tools and proteomics repositories to accommodate the extra information in a meaningful way.

So far, one attempt to mitigate those annotation problems at the technical level was the establishment of experiment and data description standard guidelines. The Minimum Information About a Microarray Experiment (MIAME) is one such format intended to provide the minimum information required to unambiguously interpret and reproduce a microarray experiment [69]. In proteomics, the Minimum Information About a Proteomics Experiment (MIAPE) guidelines [70] constitute an equivalent standard introduced by the PSI. MIAPE defines what metadata should be provided coming from whole proteomics experiments in order to be fully reproducible. Currently standards in eight proteomics subdomains are defined. The reporting guidelines for MS quantification studies were updated in 2013 [71], including now requirements for sample annotation, and are summarized in Table 1. However, to our knowledge, these guidelines are not actively implemented in any software, so they cannot be enforced in practice.

5 Conclusions

In conclusion, we note that although there is an abundance of high-quality quantitative proteomics studies in the public domain, their (re)use can still be limited due to the complexity of the proteomics workflows, heterogeneity of the underlying data, and difficulties in obtaining appropriate metadata, among other reasons. The lack of appropriate metadata is one of the more significant hurdles in our experience, especially for systematic and automated

Table 1
Minimum information about a proteomics experiment (MIAPE) guidelines

Required sections	Section description and detailed items
General features	*Overview description of the experiment:* Experiment identifier or name, responsible person or role, quantitative approach used
Experimental design and sample description	*Description of the experiment design, including treatment groups and biological/technical replication:* Sample groups, biological and technical replicates, labeling protocol (if applicable), sample description, sample name, sample amount, sample labeling with assay definition, i.e., MS run/dataset together with reporting ion mass, reagent or isotope labeled amino acid, replicates and/or groups, isotopic correction coefficients, internal references
Input data	*Description and reference of the dataset used for quantitative analysis:* Type, format, and availability of the data. No actual values are requested here: Input data type, input data format, input data merging, availability of the input data
Protocol	*Description of the software and methods applied in the quantitative analysis (including transformation functions, aggregation functions and statistical calculations):* Quantification software name, including version and manufacturer, description of the selection and/or matching method of features, confidence filter of features or peptides prior to quantification, confidence filter of features or peptides prior to quantification, description of data calculation and transformation methods, description of methods for (statistical) estimation of correctness, calibration curves of standards
Resulting data	*Quantification data should be provided here, together with relevant metrics of confidence:* Quantification values at peptide and/or feature level, quantification values at protein level

Summary of main requirements for the MIAPE MS Quantification reporting guidelines [71].

reuse. This should however not discourage future research, as successful reanalysis and data integration studies of a relatively large number of individual datasets have already been reported [31] and new computational methods will undoubtedly appear to aid in those kinds of efforts. As there is a reasonable overlap between proteomics and other omics fields where data reuse is more popular like genomics and transcriptomics, adaptations of the existing methods from those fields might prove to be a good starting point. Moreover, accessibility in public proteomics resources containing quantitative data and relevant metadata will, in our view, remain an important factor in data reuse efforts. We at PRIDE are strongly committed to facilitate and support such efforts.

Acknowledgements

The authors want to acknowledge financial support from the Wellcome Trust [grant numbers WT101477MA and 208391/Z/17/Z] and from EMBL core funds.

References

1. Larance M, Lamond AI (2015) Multidimensional proteomics for cell biology. Nat Rev Mol Cell Biol 16:269–280. https://doi.org/10.1038/nrm3970
2. Wang J, Mouradov D, Wang X et al (2017) Colorectal cancer cell line proteomes are representative of primary tumors and predict drug sensitivity. Gastroenterology 153:1082–1095. https://doi.org/10.1053/j.gastro.2017.06.008
3. Lawless C, Holman SW, Brownridge P et al (2016) Direct and absolute quantification of over 1800 yeast proteins via selected reaction monitoring. Mol Cell Proteomics 15:130–122. https://doi.org/10.1074/mcp.M115.054288
4. Lahtvee P-J, Sánchez BJ, Smialowska A et al (2017) Absolute quantification of protein and mRNA abundances demonstrate variability in gene-specific translation efficiency in yeast. Cell Syst 4:495–504.e5. https://doi.org/10.1016/j.cels.2017.03.003
5. Guo T, Kouvonen P, Koh CC et al (2015) Rapid mass spectrometric conversion of tissue biopsy samples into permanent quantitative digital proteome maps. Nat Med 21:407–413
6. Kulak NA, Pichler G, Paron I et al (2014) Minimal, encapsulated proteomic-sample processing applied to copy-number estimation in eukaryotic cells. Nat Methods 11:319–324. https://doi.org/10.1038/nmeth.2834
7. Navarro P, Kuharev J, Gillet LC et al (2016) A multicenter study benchmarks software tools for label-free proteome quantification. Nat Biotechnol 34:1130–1136. https://doi.org/10.1038/nbt.3685
8. Tyanova S, Temu T, Cox J (2016) The MaxQuant computational platform for mass spectrometry-based shotgun proteomics. Nat Protoc 11:2301–2319. https://doi.org/10.1038/nprot.2016.136
9. Hebert AS, Richards AL, Bailey DJ et al (2014) The one hour yeast proteome. Mol Cell Proteomics 13:339–347. https://doi.org/10.1074/mcp.M113.034769
10. Perry RH, Cooks RG, Noll RJ (2008) Orbitrap mass spectrometry: instrumentation, ion motion and applications. Mass Spectrom Rev 27:661–699. https://doi.org/10.1002/mas.20186
11. Vizcaíno JA, Csordas A, del-Toro N et al (2016) 2016 update of the PRIDE database and its related tools. Nucleic Acids Res 44:D447–D456. https://doi.org/10.1093/nar/gkv1145

12. Martens L, Hermjakob H, Jones P et al (2005) PRIDE: the proteomics identifications database. Proteomics 5:3537–3545. https://doi.org/10.1002/pmic.200401303
13. Deutsch EW, Csordas A, Sun Z et al (2017) The ProteomeXchange consortium in 2017: supporting the cultural change in proteomics public data deposition. Nucleic Acids Res 45:D1100–D1106. https://doi.org/10.1093/nar/gkw936
14. Vizcaíno JA, Deutsch EW, Wang R et al (2014) ProteomeXchange provides globally coordinated proteomics data submission and dissemination. Nat Biotechnol 32:223–226. https://doi.org/10.1038/nbt.2839
15. Okuda S, Watanabe Y, Moriya Y et al (2017) jPOSTrepo: an international standard data repository for proteomes. Nucleic Acids Res 45:D1107–D1111. https://doi.org/10.1093/nar/gkw1080
16. Vaudel M, Verheggen K, Csordas A et al (2016) Exploring the potential of public proteomics data. Proteomics 16:214–225. https://doi.org/10.1002/pmic.201500295
17. Martens L, Vizcaíno JA (2017) A golden age for working with public proteomics data. Trends Biochem Sci 42:333–341. https://doi.org/10.1016/j.tibs.2017.01.001
18. Perez-Riverol Y, Alpi E, Wang R et al (2015) Making proteomics data accessible and reusable: current state of proteomics databases and repositories. Proteomics 15:930–949. https://doi.org/10.1002/pmic.201400302
19. Craig R, Cortens JP, Beavis RC (2004) Open source system for analyzing, validating, and storing protein identification data. J Proteome Res 3:1234–1242. https://doi.org/10.1021/PR049882H
20. Desiere F, Deutsch EW, King NL et al (2006) The PeptideAtlas project. Nucleic Acids Res 34:D655–D658. https://doi.org/10.1093/nar/gkj040
21. Farrah T, Deutsch EW, Kreisberg R et al (2012) PASSEL: the PeptideAtlas SRMexperiment library. Proteomics 12:1170–1175. https://doi.org/10.1002/pmic.201100515
22. Jones AR, Eisenacher M, Mayer G et al (2012) The mzIdentML data standard for mass spectrometry-based proteomics results. Mol Cell Proteomics 11:M111.014381. https://doi.org/10.1074/mcp.M111.014381
23. Griss J, Jones AR, Sachsenberg T et al (2014) The mzTab data exchange format: communicating mass-spectrometry-based proteomics and metabolomics experimental results to a wider audience. Mol Cell Proteomics 13:2765–2775. https://doi.org/10.1074/mcp.O113.036681
24. Martens L, Chambers M, Sturm M et al (2011) mzML—a community standard for mass spectrometry data. Mol Cell Proteomics 10:R110.000133. https://doi.org/10.1074/mcp.R110.000133
25. Perez-Riverol Y, Xu Q-W, Wang R et al (2016) PRIDE Inspector toolsuite: moving toward a universal visualization tool for proteomics data standard formats and quality assessment of ProteomeXchange datasets. Mol Cell Proteomics 15:305–317. https://doi.org/10.1074/mcp.O115.050229
26. Ellis MJ, Gillette M, Carr SA et al (2013) Connecting genomic alterations to cancer biology with proteomics: the NCI Clinical Proteomic Tumor Analysis Consortium. Cancer Discov 3:1108–1112. https://doi.org/10.1158/2159-8290.CD-13-0219
27. Zhang H, Liu T, Zhang Z et al (2016) Integrated proteogenomic characterization of human high-grade serous ovarian cancer. Cell 166:755–765. https://doi.org/10.1016/j.cell.2016.05.069
28. Mertins P, Mani DR, Ruggles KV et al (2016) Proteogenomics connects somatic mutations to signalling in breast cancer. Nature 534:55–62
29. Rudnick PA, Markey SP, Roth J et al (2016) A description of the Clinical Proteomic Tumor Analysis Consortium (CPTAC) common data analysis pipeline. J Proteome Res 15:1023–1032. https://doi.org/10.1021/acs.jproteome.5b01091
30. Edwards NJ, Oberti M, Thangudu RR et al (2015) The CPTAC data portal: a resource for cancer proteomics research. J Proteome Res 14:2707–2713. https://doi.org/10.1021/pr501254j
31. Wilhelm M, Schlegl J, Hahne H et al (2014) Mass-spectrometry-based draft of the human proteome. Nature 509:582–587
32. Zolg DP, Wilhelm M, Schnatbaum K et al (2017) Building ProteomeTools based on a complete synthetic human proteome. Nat Meth 14:259–262
33. Perkins DN, Pappin DJC, Creasy DM, Cottrell JS (1999) Probability-based protein identification by searching sequence databases using mass spectrometry data. Electrophoresis 20:3551–3567. https://doi.org/10.1002/(SICI)1522-2683(19991201)20:18<3551::AID-ELPS3551>3.0.CO;2-2
34. Fagerberg L, Hallström BM, Oksvold P et al (2014) Analysis of the human tissue-specific expression by genome-wide integration of transcriptomics and antibody-based proteomics. Mol Cell Proteomics 13:397–406. https://doi.org/10.1074/mcp.M113.035600

35. Kim M-S, Pinto SM, Getnet D et al (2014) A draft map of the human proteome. Nature 509:575–581
36. Eng JK, McCormack AL, Yates JR (1994) An approach to correlate tandem mass spectral data of peptides with amino acid sequences in a protein database. J Am Soc Mass Spectrom 5:976–989. https://doi.org/10.1016/1044-0305(94)80016-2
37. Wang M, Weiss M, Simonovic M et al (2012) PaxDb, a database of protein abundance averages across all three domains of life. Mol Cell Proteomics 11:492–500. https://doi.org/10.1074/mcp.O111.014704
38. Szklarczyk D, Franceschini A, Wyder S et al (2015) STRING v10: protein-protein interaction networks, integrated over the tree of life. Nucleic Acids Res 43:D447–D452. https://doi.org/10.1093/nar/gku1003
39. Wang M, Herrmann CJ, Simonovic M et al (2015) Version 4.0 of PaxDb: protein abundance data, integrated across model organisms, tissues, and cell-lines. Proteomics 15:3163–3168. https://doi.org/10.1002/pmic.201400441
40. Schaab C, Geiger T, Stoehr G et al (2012) Analysis of high accuracy, quantitative proteomics data in the MaxQB database. Mol Cell Proteomics 11:M111.014068. https://doi.org/10.1074/mcp.M111.014068
41. Cox J, Mann M (2008) MaxQuant enables high peptide identification rates, individualized p.p.b.-range mass accuracies and proteome-wide protein quantification. Nat Biotechnol 26:1367–1372. https://doi.org/10.1038/nbt.1511
42. Conesa A, Madrigal P, Tarazona S et al (2016) A survey of best practices for RNA-seq data analysis. Genome Biol 17:13. https://doi.org/10.1186/s13059-016-0881-8
43. Bittremieux W, Meysman P, Martens L et al (2016) Unsupervised quality assessment of mass spectrometry proteomics experiments by multivariate quality control metrics. J Proteome Res 15:1300–1307. https://doi.org/10.1021/acs.jproteome.6b00028
44. Bittremieux W, Walzer M, Tenzer S et al (2017) The human proteome organization-proteomics standards initiative quality control working group: making quality control more accessible for biological mass spectrometry. Anal Chem 89:4474–4479. https://doi.org/10.1021/acs.analchem.6b04310
45. Bantscheff M, Lemeer S, Savitski MM, Kuster B (2012) Quantitative mass spectrometry in proteomics: critical review update from 2007 to the present. Anal Bioanal Chem 404:939–965. https://doi.org/10.1007/s00216-012-6203-4
46. Domon B, Aebersold R (2010) Options and considerations when selecting a quantitative proteomics strategy. Nat Biotechnol 28:710–721
47. Shi T, Song E, Nie S et al (2016) Advances in targeted proteomics and applications to biomedical research. Proteomics 16:2160–2182. https://doi.org/10.1002/pmic.201500449
48. Hu A, Noble WS, Wolf-Yadlin A (2016) Technical advances in proteomics: new developments in data-independent acquisition. F1000Res 5. pii: F1000 Faculty Rev-419. https://doi.org/10.12688/f1000research.7042.1
49. Yates JR, Ruse CI, Nakorchevsky A (2009) Proteomics by mass spectrometry: approaches, advances, and applications. Annu Rev Biomed Eng 11:49–79. https://doi.org/10.1146/annurev-bioeng-061008-124934
50. Tu C, Sheng Q, Li J et al (2015) Optimization of search engines and postprocessing approaches to maximize peptide and protein identification for high-resolution mass data. J Proteome Res 14:4662–4673. https://doi.org/10.1021/acs.jproteome.5b00536
51. Shteynberg D, Nesvizhskii AI, Moritz RL, Deutsch EW (2013) Combining results of multiple search engines in proteomics. Mol Cell Proteomics 12:2383–2393. https://doi.org/10.1074/mcp.R113.027797
52. Ting L, Cowley MJ, Hoon SL et al (2009) Normalization and statistical analysis of quantitative proteomics data generated by metabolic labeling. Mol Cell Proteomics 8:2227–2242. https://doi.org/10.1074/mcp.M800462-MCP200
53. Karpievitch YV, Dabney AR, Smith RD (2012) Normalization and missing value imputation for label-free LC-MS analysis. BMC Bioinformatics 13(Suppl 16):S5. https://doi.org/10.1186/1471-2105-13-S16-S5
54. Välikangas T, Suomi T, Elo LL et al (2016) A systematic evaluation of normalization methods in quantitative label-free proteomics. Brief Bioinform 86:bbw095. https://doi.org/10.1093/bib/bbw095
55. Arike L, Valgepea K, Peil L et al (2012) Comparison and applications of label-free absolute proteome quantification methods on *Escherichia coli*. J Proteome 75:5437–5448. https://doi.org/10.1016/j.jprot.2012.06.020
56. Taverner T, Karpievitch YV, Polpitiya AD et al (2012) DanteR: an extensible R-based tool for quantitative analysis of -omics data. Bioinformatics 28:2404–2406. https://doi.org/10.1093/bioinformatics/bts449

57. Chawade A, Alexandersson E, Levander F (2014) Normalyzer: a tool for rapid evaluation of normalization methods for omics data sets. J Proteome Res 13:3114–3120. https://doi.org/10.1021/pr401264n
58. Pedrioli PGA, Eng JK, Hubley R et al (2004) A common open representation of mass spectrometry data and its application to proteomics research. Nat Biotechnol 22:1459–1466. https://doi.org/10.1038/nbt1031
59. Chambers MC, Maclean B, Burke R et al (2012) A cross-platform toolkit for mass spectrometry and proteomics. Nat Biotechnol 30:918–920. https://doi.org/10.1038/nbt.2377
60. Kessner D, Chambers M, Burke R et al (2008) ProteoWizard: open source software for rapid proteomics tools development. Bioinformatics 24:2534–2536. https://doi.org/10.1093/bioinformatics/btn323
61. Perez-Riverol Y, Wang R, Hermjakob H et al (2014) Open source libraries and frameworks for mass spectrometry based proteomics: a developer's perspective. Biochim Biophys Acta 1844:63–76. https://doi.org/10.1016/j.bbapap.2013.02.032
62. Walzer M, Qi D, Mayer G et al (2013) The mzQuantML data standard for mass spectrometry-based quantitative studies in proteomics. Mol Cell Proteomics 12:2332–2340. https://doi.org/10.1074/mcp.O113.028506
63. Jarnuczak AF, Lee DCH, Lawless C et al (2016) Analysis of intrinsic peptide detectability via integrated label-free and SRM-based absolute quantitative proteomics. J Proteome Res 15:2945–2959. https://doi.org/10.1021/acs.jproteome.6b00048
64. Falick AM, Lane WS, Lilley KS et al (2011) ABRF-PRG07: advanced quantitative proteomics study. J Biomol Tech 22:21–26
65. Selevsek N, Chang C-Y, Gillet LC et al (2015) Reproducible and consistent quantification of the *Saccharomyces cerevisiae* proteome by SWATH-mass spectrometry. Mol Cell Proteomics 14:739–749. https://doi.org/10.1074/mcp.M113.035550
66. Lee MV, Topper SE, Hubler SL et al (2011) A dynamic model of proteome changes reveals new roles for transcript alteration in yeast. Mol Syst Biol 7:514. https://doi.org/10.1038/msb.2011.48
67. Goveia J, Pircher A, Conradi L et al (2016) Meta-analysis of clinical metabolic profiling studies in cancer: challenges and opportunities. EMBO Mol Med 8:1134–1142
68. Griss J, Perez-Riverol Y, Hermjakob H, Vizcaíno JA (2015) Identifying novel biomarkers through data mining—a realistic scenario? Proteomics Clin Appl 9:437–443. https://doi.org/10.1002/prca.201400107
69. Brazma A, Hingamp P, Quackenbush J et al (2001) Minimum information about a microarray experiment (MIAME)-toward standards for microarray data. Nat Genet 29:365–371. https://doi.org/10.1038/ng1201-365
70. Taylor CF, Paton NW, Lilley KS et al (2007) The minimum information about a proteomics experiment (MIAPE). Nat Biotechnol 25:887–893. https://doi.org/10.1038/nbt1329
71. Martínez-Bartolomé S, Deutsch EW, Binz P-A et al (2013) Guidelines for reporting quantitative mass spectrometry based experiments in proteomics. J Proteome 95:84–88. https://doi.org/10.1016/j.jprot.2013.02.026

Chapter 15

Computational Proteomics with Jupyter and Python

Lars Malmström

Abstract

Proteomics based on mass spectrometry produces complex data in large quantities. The need for flexible computational pipelines, in the context of big data, in proteomics and other areas of science, has prompted the development of computational platforms and libraries that facilitate data analysis and data processing. In this respect, Python appears to be one of the winners among programming languages in terms of popularity and development. This chapter shows how to perform basic tasks using Python and dedicated libraries in a Jupyter framework: from basic search result summarizations to the creation of MS1 chromatograms.

Key words Proteomics, Python, R, Jupyter, JupyterHub, Reproducible research

1 Introduction

Mass spectrometry-based proteomics is a powerful technology. Three aspects that set it apart are speed, flexibility, and its applicability to arbitrarily complex samples. Ion optics and high-resolution mass analyzers enable the user to measure anything that alters the chemical composition of the molecules of interest, at least in principle. Scientists use this ability to identify and quantify many aspects of their samples that are otherwise difficult to study, such as post-translational modifications, or they use chemistry to identify amino acids that are surface exposed or measure protein-protein interactions. As a result, the list of protocols and data acquisition methods is long; while this is good, it also causes complexity of the software used to analyze the data that results from these experiments. On the one hand, we now have powerful software suits that are desktop-based and, to some degree, are relatively easy to use; on the other hand, software suits exist that were designed bottom up to deal with common protocols or very large experiments where a single computer simply is not powerful enough to process the data in a reasonable time. There are many use cases that fall in between these two extremes, where researchers want to prototype a

Caroline A. Evans et al. (eds.), *Mass Spectrometry of Proteins: Methods and Protocols*, Methods in Molecular Biology, vol. 1977,
https://doi.org/10.1007/978-1-4939-9232-4_15,

workflow to analyze a new type of data or to study an interesting or unusual aspect of a given dataset. The standardized solutions are simply not flexible enough and writing software from scratch is not an option due to time constraints or skill gaps. In this book chapter, I will show a few examples of how mass spectrometrists with some basic programming skills can take advantage of the many powerful python libraries to explore their data. To allow the reader to get some practical experience and to run some of these examples themselves, with a minimal amount of time spent on installing software and downloading data, I provide a Docker container with needed software installed. All examples are demonstrated using Jupyter Notebooks which also allow for relatively simple integration of JavaScript, making the integration of tools such as Bokeh for interactive visualization. This chapter has two main sections, an introduction to Docker containers and Jupyter, with specific instructions on how to get the Docker container accompanying this chapter running. The second section of the chapter will walk through a few software packages and examples of how to extract and visualize data using Python. Please feel free to skip down to Python in Proteomics if you are not interested in the technical aspects.

2 Materials

2.1 Instructions to Install Docker Containers

1. Mac: https://docs.docker.com/docker-for-mac/install/.
2. Windows 10 Pro: https://docs.docker.com/docker-for-windows/install/.
3. Older versions of Windows: https://docs.docker.com/toolbox/toolbox_install_windows/.
4. Linux servers: https://docs.docker.com/engine/installation/.
5. Python https://www.python.org/ (*see* **Note 1**).
6. The pandas library: https://pandas.pydata.org/ (*see* **Note 1**).
7. The NumPy library: http://www.numpy.org/ (*see* **Note 1**).
8. The Ursgal library: https://github.com/ursgal/ursgal.
9. The Bokeh library: https://bokeh.pydata.org/.
10. The PyLab library: https://scipy.github.io/old-wiki/pages/PyLab.
11. The Pyteomics library: https://pypi.org/project/pyteomics/.
12. The PyOpenMS library: https://github.com/OpenMS/OpenMS/wiki/pyOpenMS.
13. Jupyter: http://jupyter.org/ (*see* **Note 1**).

3 Methods

3.1 JupyterHub as a Docker Container

Docker containers are to software what shipping containers are to goods—a number of different computers/operating systems can run various tool packaged in a standardized "box." Developers can package several different software packages into a Docker container, and these can then be executed on computers of different sizes and host operating systems. The technologies on which this chapter depends were developed with servers in mind, and this is, of course, great as servers are larger and hence can handle the workload better. However, the overhead of setting up and running servers is too high for somebody to try something out or to run a tutorial as in this example. The second advantage of Docker containers is that they can be versioned, and this means you can go back to a previous version if needed. The only drawback is that installing the Docker software that run Docker containers requires a bit of effort, especially if you are using the Windows operating system in any other version that Windows 10 Professional (although Microsoft might have mitigated this shortcoming by the time this gets to press). *See* links in the Materials section to find excellent tutorials on how to install docker on the most common operating systems. Docker also has an excellent community that can support you with any issue you encounter. For the example provided here, configure Docker to use 6GB of RAM or more.

The concept of Docker is shown in Fig. 1. A developer builds a container (orange box) on his or her computer, shown in red. That container is then published on, for example, Docker Hub, where each version of that container is available for download. Users using different operating systems, Windows (blue), Mac (dark green), or Linux (light green) for example, can pull the container from Docker Hub.

1. Download the container from Docker Hub on the command line using
```
docker pull malmstroem/pyproteomics
```
2. Start the container, listen to port 8000, run the command jupyterhub with the command (make sure you replace the string [*host folder*] with a folder on your host computer)
```
docker run --name jhub -p 8000:8000
-v [host folder]:/host malmstroem/
pyproteomics:latest jupyterhub
```
   The -v flag makes a disk or part of a disk available inside the container. The -p flag makes it possible for a so-called port of the container to be linked to a port of the host. Here, the container port 8000 is mapped to the host port 8000 and this allows direct interaction with the program inside the container that is configured to communicate over port 8000 (*see* **Note 2**).

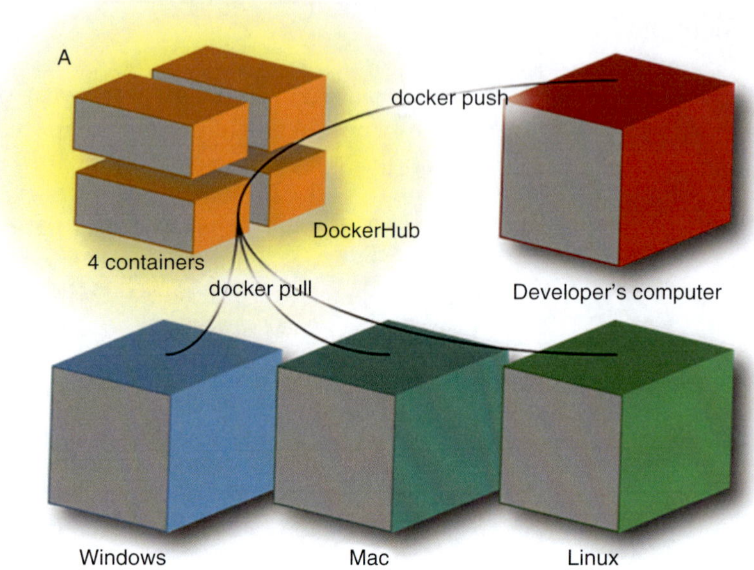

Fig. 1 General schematic illustrating the principle of Docker containers

3. Load the page http://127.0.0.1:8000 in your browser and log in using "user" and "password" as the user name and password.
4. Click on the demo notebook to get started.

3.2 Getting Started with Jupyter Notebooks

Jupyter is a technology that enables users to execute code (such as Python or R) on a remote server, using a rich web application. Jupyter is organized around documents called "notebooks" and notebooks in the current working directory are displayed in a table format in the main JupyterHub page, *see* Fig. 2a. There are two important functions available as buttons in the right part of the interface: "Upload" and "New." "Upload" allows the user to upload any file from their computer to the server and "New" displays a menu giving the user the opportunity to create a new notebook (several programming languages, referred to as "kernels," are generally available), to create a text file, to create a new folder, or to launch a shell (*see* **Note 3**).

1. Create a new Python 3 notebook by clicking "New" and by selecting "Python 3." This creates a Python 3 notebook in a new tab of the web browser; by default, the notebook is named "Untitled."
2. Rename the notebook by clicking on "Untitled."
3. Click on the first cell and type

```
a = 5
b = 3
```

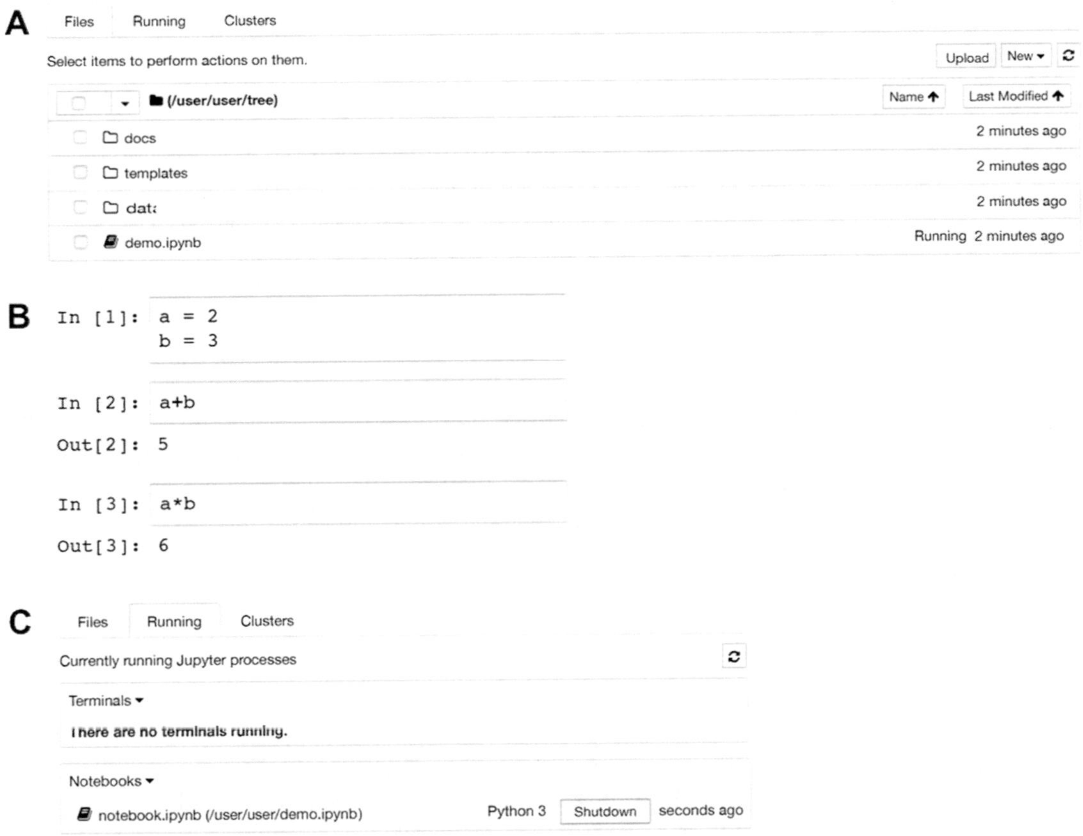

A

Files Running Clusters

Select items to perform actions on them.

Upload New ▾

(/user/user/tree) Name ↑ Last Modified ↑

docs 2 minutes ago

templates 2 minutes ago

dat: 2 minutes ago

demo.ipynb Running 2 minutes ago

B

```
In [1]: a = 2
        b = 3

In [2]: a+b

Out[2]: 5

In [3]: a*b

Out[3]: 6
```

C

Files Running Clusters

Currently running Jupyter processes

Terminals ▾

There are no terminals running.

Notebooks ▾

notebook.ipynb (/user/user/demo.ipynb) Python 3 Shutdown seconds ago

Fig. 2 (**a**) The main page of the Jupyter interface lists notebooks, files, and folders in the main panel. Above the main panel are options to see running notebooks and terminals as well as buttons to upload files, and create new notebooks, terminals, folders, and files. (**b**) The notebook consists of cells, marked with In[] and a number in the bracket that corresponds to how many cells have been executed in the notebook. The output of the cell is found below, marked with Out[]. (**c**) In the "Running" tab, terminals and notebooks that are running on the server are listed. They can be shut down to free up resources on the server

Then simultaneously hit SHIFT and ENTER to evaluate the cell; the cell should get a label "In [1]" to indicate that it's been sent to the Python kernel for processing. As these are two variable assignments, no output is produced.

4. The cursor should now be in a new cell waiting for more input; type

```
a + b
```

then hit the SHIFT and ENTER keys to evaluate the cell. The cell gets labeled "In [2]" and produces an output that is labeled "Out [2]."

5. The cursor should now be in a new cell waiting for more input; type

```
a * b
```

6. Then hit the SHIFT and ENTER keys to evaluate the cell. The cell gets labeled "In [3]" and produces an output that is labeled "Out [3]" (Fig. 2b).
7. Go back the previous tab in the web browser; the notebook now appears in the list of files and Jupyter shows that the Python 3 kernel is running. Alternatively, click on the "Running" tab in the Jupyter page to check what's going on (Fig. 2c). You can kill the kernel but clicking on "Shutdown" although it's generally best to halt the kernel directly from the notebook by selecting "Close and Halt" in the "File" menu.

3.3 Getting Started with Python Programming

Python is a programming language that is both powerful and simple to use; it has gained popularity in the bioinformatics, proteomics, and machine learning communities. It is beyond the scope of this chapter to talk in depth about Python and a large number of software packages that are installed. I'll illustrate the use of a standard library. The library "pandas" allows Python to read and manipulate data in a similar ways to Excel can manipulate data in a spreadsheet. I'll illustrate here how pandas can be used to manipulate the peptide-spectrum matches (PSMs) contained in the "protein_peptide.xlsx" file.

1. Create a new Python 3 notebook by clicking "New" and by selecting "Python 3." This creates a Python 3 notebook in a new tab of the web browser; by default, the notebook is named "Untitled." Rename the notebook by clicking on "Untitled."
2. In a Python cell, import the library "pandas" (conventionally, for legibility purposes, the library prefix imported under the "pd" name)

```
import pandas as pd
```

3. Import the PSMs

```
df = pd.read_excel("protein_peptide.xlsx", 0)
```

   This assigns the data as a data frame to the variable "df." A data frame is an array whose columns has names.

4. You can get an idea of the contents of the data by looking at a few entries

```
df.head()
```

5. Alternatively, you can focus on one specific column using

```
df.peptide.head()
```

   Here "df.peptide" refers to the column "peptide" of the data frame "df."

6. You can summarize the number of peptides by applying the methods groupby() and count() to the data frame to obtain the final pivot table

```
df.groupby("protein").count().
```

3.4 Searching a Database

In this section, I intend to demonstrate a small subset of all the powerful Python packages that exist. This is by no means comprehensive, but I hope to show how to process, access and visualize data and results from proteomics workflows. The data was produced using a Thermo QExactive+ to measure mouse blood plasma [1]. For convenience, all commands mentioned below are provided in the "demo" notebook in the Docker container.

1. In a new notebook, import libraries [2]

```
import ursgal
import pandas as pd
```

2. Create a UController class and perform a search using the database "mouse_panther12.fasta" and the spectra contained in "Erik_S1407_239.mzML" [3, 4]

```
uc = ursgal.UController(
params = {'database': 'mouse_panther12.
fasta'},
profile = 'QExactive+')
result = uc.search(input_file = 'Erik_
S1407_239.mzML',
engine = 'xtandem_vengeance')
```

   This will produce a CSV file containing the PSMs.

3. The CSV file can be parsed and analyzed using pandas

```
unif = pd.read_csv('Erik_S1407_239_xtandem_
vengeance_pmap_unified.csv')
```

   For instance, we can get the dimensions of the data using

```
unif.shape[1]
```

   What columns are present in the file (Spectrum ID, Spectrum Title, Expected m/z, etc.) can be obtained

```
unif.columns
```

   A given column can be used a separate entity using something like

```
unif['Mass Difference']
```

4. The demo notebook also includes the code that allows one to plot the histogram of X!Tandem *e*-values contained in the "X!Tandem:expect" column of the data frame "unif." High-confidence PSMs can thus be selected from the dataset using

```
hi_conf = unif[unif['X\!Tandem:expect'] <
2.382e-15]
```

   The innermost unif instruction creates a list of booleans that is used by the outermost unif instruction to select the search results in the data frame such that the *e*-value is less than 2.382×10^{-15}.

3.5 Spectrum Visualization

Using pandas, one can easily find, among the high-confidence identifications, the most common peptide, RHPDYSVSLLLR from Mouse Serum Albumin. That step is described in the demo notebook. Here, we set about visualizing and visually validating one spectrum assigned to that peptide.

1. First let's assign the peptide of interest's sequence and the charge state we're interested in

```
peptide = 'RHPDYSVSLLLR'
charge = 3
```

2. Import a few libraries

```
from pyteomics import mgf, mass
import pylab
```

3. We define a function whose purpose is to compute all theoretical fragments produced by the fragmentation of a peptide (*see* **Note 4**)

```
     def fragments(peptide): # ,
types=('b', 'y'), maxcharge=1
          """ This function calculates frag-
ments for a given peptide """
          for I in range(1, len(peptide)-1):
               for ion_type in ('b','y'):
                    if ion_type[0] in 'abc':
                         yield mass.fast_
mass(peptide[:i], ion_type=ion_type,
charge=1)
                    else:
                         yield mass.fast_
mass(peptide[i:], ion_type=ion_type,
charge=1)
```

4. Let's read the first and only spectrum from the file "Erik_S1407_239.25966.25966.3.mgf" (MGF stands for Mascot Generic File)

```
  with mgf.read('Erik_
S1407_239.25966.25966.3.mgf') as mgf_file:
    spectrum = next(mgf_file)
```

The variable "spectrum" is what is technically known as a dictionary with four keys: "*m/z* array," "intensity array," "charge array," and "params." We'll use those when producing the final plot; "spectrum['params']" is itself a dictionary that gives access to the charge state (key "charge"), the precursor mass ("pepmass"), the retention time ("rtinseconds"), and the so-called spectrum title ("title"), among other things (*see* **Note 5**).

5. Let's now plot the spectrum. A wider format is used for the spectrum representation:

```
pylab.figure(figsize=(15,5))
```

The plot is given a title

```
pylab.title('Spectrum Erik_
S1407_239.25966.25966.3 with theoretical
fragments of ' + peptide)
```

Axes are labeled

```
pylab.xlabel('m/z')
pylab.ylabel('Intensity')
```

We compute the theoretical fragments of the peptides

```
theor_spectrum = list(fragments(peptide))
```

First red bars are placed where the theoretical fragments are expected:

```
pylab.bar(theor_spectrum,
[spectrum['intensity array'].max()] *
len(theor_spectrum),
     width=0.1, edgecolor='red', alpha=0.7)
```

Then the observed spectrum is added as back bars:

```
pylab.bar(spectrum['m/z array'],
     spectrum['intensity array'],
     width=0.1, linewidth=2,
edgecolor='black')
```

The final plot gets shown (Fig. 3) using

```
pylab.show()
```

3.6 MS1 Chromatogram

Here we take a closer look at the MS1 chromatogram for the same peptide RHPDYSVSLLLR. To do this, we can explicitly write some code to create the chromatogram with the parameters we want. There are of course software packages such as OpenSWATH that can accomplish this much more efficiently [5]. For this, I am here using PyOpenMS [6], the Python binding for the large, fast, and versatile C++ software suit called OpenMS [7], something that can be useful in case NumPress [8] was used to reduce the size of the mzML file. PyOpenMS is only available in Python 2, but as stated above, we can simply execute this in a Python 2 kernel by including "%%python2" at the top of the cell. (Another approach is also illustrated in the demo notebook.)

1. In a Python 2 cell, load the necessary libraries

```
%%python2
import pyopenms
import pandas as pd
```

Import the data

```
msExp = pyopenms.MSExperiment()
inputfile = 'Erik_S1407_239.mzML'
pyopenms.FileHandler().
loadExperiment(inputfile, msExp)
```

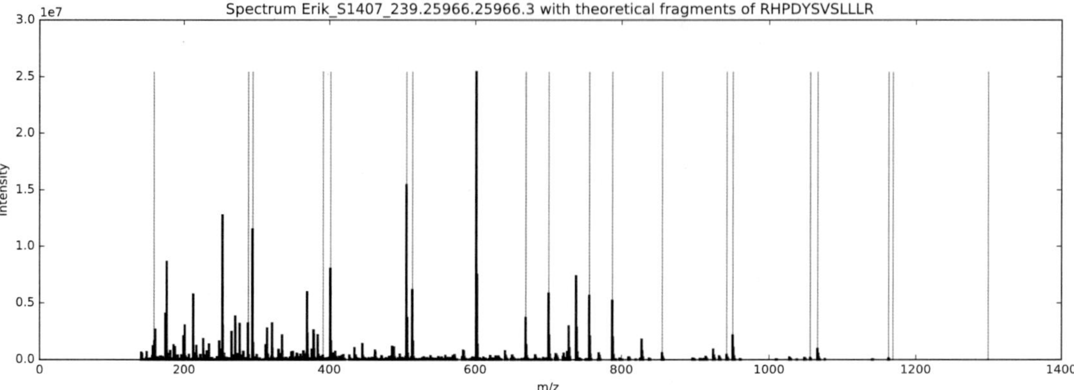

Fig. 3 MS2 spectrum overlaid with the masses of the theoretical fragments of peptide RHPDYSVSLLLR for expect inspection

Collect all peaks in MS1 spectra that have a mass equal to 485.94045, the mono-isotopic mass of the peptide of interest when it has a 3+ charge (*see* **Note 6**):

```
entries = []
for spectrum in msExp:
    if not spectrum.getMSLevel() == 1:
        continue
    for peak in spectrum:
        if abs(peak.getMZ() - 485.94045) >
0.01:
            continue
        entries.append({'rt': spectrum.
getRT(), 'mz': peak.getIntensity()})
```

The resulting data are saved as a tab-separated file

```
chromtable = pd.DataFrame(entries)
chromtable.to_csv('ms1chromatogram.tsv',
                  index=False, sep="\t")
```

2. The chromatogram needs to be processed to be of any use. One way to do this is to bin the peaks by retention time and to only keep the maximum intensity for each bin. In a new (Python 3) cell, we can write the following.

   Import the NumPy library

```
import numpy as np
```

   Import the data as a dataframe

```
ms1 = pd.read_table('ms1chromatogram.tsv')
```

   The "delta" variable is the size of the retention bin.

```
delta = 25
```

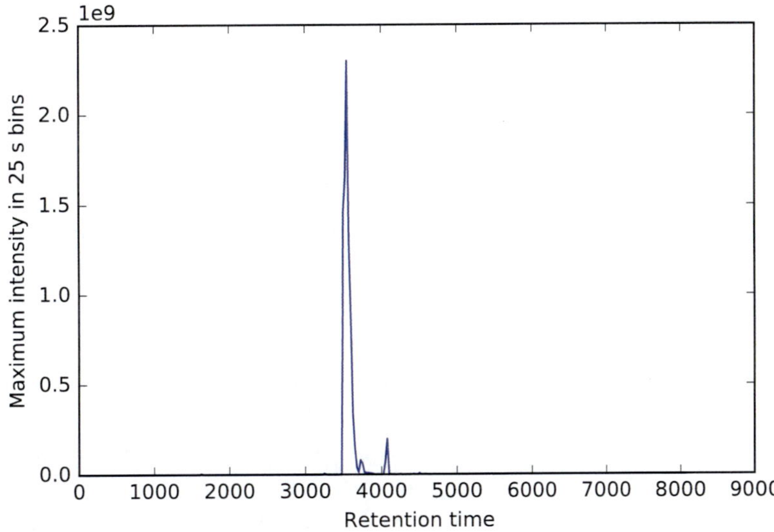

Fig. 4 MS1 chromatogram for the same peptide RHPDYSVSLLLR

Initialize the chromatogram

```
imax = int(ms1['rt'].max() // delta)
chromato = np.array([[rt * delta, 0.] for
rt. in range(imax + 1)])
```

Read the data and update the chromatogram

```
for index, row in ms1.iterrows():
    rt = row['rt']
    mz = row['mz']
    i = int(rt // delta)
    chromato[i][1] = max(chromato[i] [1], mz)
```

3. The visualization can be achieved using PyLab (Fig. 4)

```
pylab.plot(chromato[:, 0], chromato[:, 1])
pylab.xlabel("Retention time")
pylab.ylabel("Maximum intensity in 25 s bins")
pylab.show()
```

3.7 Stopping the Docker Container

1. Stop the docker with the following command

```
docker stop jhub
```

2. Delete the container and all its files with

```
docker rm jhub
```

4 Notes

1. Conda is an open source package management system for Windows, macOS, and Linux, which allows the user to quickly install and update packages for Python among other things. It includes Jupyter but unfortunately the ursgal library is not included; URL: https://conda.io/docs/index.html.
2. Another way to interact with the Docker container is to issue a "docker exec" command in the shell to execute a command inside the Docker container.
3. Useful tips are available from https://www.dataquest.io/blog/jupyter-notebook-tips-tricks-shortcuts/.
4. The example in this chapter specifically looks at B and Y ion series; for other series, the reader can have a look at the Pyteomics documentation https://pyteomics.readthedocs.io/en/latest/examples/example_msms.html.
5. If one wanted to iterate through all spectra of an MGF file, you could do as follows:
```
for spectrum in mgf.read(mgf_name):
    # process spectrum.
```
6. One can for instance calculate this mass using the "Fragment Ion Calculator" http://db.systemsbiology.net:8080/proteomicsToolkit/FragIonServlet.html.

References

1. Malmström E, Kilsgård O, Hauri S et al (2016) Large-scale inference of protein tissue origin in gram-positive sepsis plasma using quantitative targeted proteomics. Nat Commun 7:10261
2. Kremer LPM, Leufken J, Oyunchimeg P et al (2016) Ursgal, universal Python module combining common bottom-up proteomics tools for large-scale analysis. J Proteome Res 15:788–794
3. Mi H, Huang X, Muruganujan A et al (2017) PANTHER version 11: expanded annotation data from Gene Ontology and Reactome pathways, and data analysis tool enhancements. Nucleic Acids Res 45:D183–D189
4. Craig R, Beavis RC (2004) TANDEM: matching proteins with tandem mass spectra. Bioinformatics 20:1466–1467
5. Röst HL, Rosenberger G, Navarro P et al (2014) OpenSWATH enables automated, targeted analysis of data-independent acquisition MS data. Nat Biotechnol 32:219–223
6. Röst HL, Schmitt U, Aebersold R, Malmstrom L (2014) pyOpenMS: a Python-based interface to the OpenMS mass-spectrometry algorithm library. Proteomics 14:74–77
7. Röst HL, Sachsenberg T, Aiche S et al (2016) OpenMS: a flexible open-source software platform for mass spectrometry data analysis. Nat Methods 13:741–748
8. Teleman J, Dowsey AW, Gonzalez-Galarza FF et al (2014) Numerical compression schemes for proteomics mass spectrometry data. Mol Cell Proteomics 13:1537–1542

Chapter 16

The Galaxy Platform for Reproducible Affinity Proteomic Mass Spectrometry Data Analysis

Paul A. Stewart, Brent M. Kuenzi, Subina Mehta, Praveen Kumar, James E. Johnson, Pratik Jagtap, Timothy J. Griffin, and Eric B. Haura

Abstract

Affinity proteomics (AP-MS) is growing in importance for characterizing protein-protein interactions (PPIs) in the form of protein complexes and signaling networks. The AP-MS approach necessitates several different software tools, integrated into reproducible and accessible workflows. However, if the scientist (e.g., a bench biologist) lacks a computational background, then managing large AP-MS datasets can be challenging, manually formatting AP-MS data for input into analysis software can be error-prone, and data visualization involving dozens of variables can be laborious. One solution to address these issues is Galaxy, an open source and web-based platform for developing and deploying user-friendly computational pipelines or workflows. Here, we describe a Galaxy-based platform enabling AP-MS analysis. This platform enables researchers with no prior computational experience to begin with data from a mass spectrometer (e.g., peaklists in mzML format) and perform peak processing, database searching, assignment of interaction confidence scores, and data visualization with a few clicks of a mouse. We provide sample data and a sample workflow with step-by-step instructions to quickly acquaint users with the process.

Key words Affinity purification, Affinity proteomics, APOSTL, AP-MS, Galaxy-P

1 Introduction

Protein-protein interactions (PPIs) are a fundamental type of physical interaction between two or more proteins [1]. These interactions are important as they give rise to nearly all biological processes. Alterations in native PPIs can be cause or consequence of human disease, and as such, they are increasingly being seen as promising drug targets [2]. Since proteins are almost always part of larger protein complexes and complex signaling networks, robust and reproducible methods are necessary for facilitating our understanding of PPIs. Further, due to the breadth and scope of possible interactions, methods that are capable of simultaneously detecting multiple proteins are often required.

Caroline A. Evans et al. (eds.), *Mass Spectrometry of Proteins: Methods and Protocols*, Methods in Molecular Biology, vol. 1977, https://doi.org/10.1007/978-1-4939-9232-4_16, © Springer Science+Business Media, LLC, part of Springer Nature 2019

Affinity proteomics (AP-MS) has become a popular experimental method for reliably querying the universe of PPIs, often referred to as the interactome [3]. A typical AP-MS workflow consists of the immunoprecipitation (IP) or "pull-down" of a target protein of interest using an antibody followed by injection into a liquid chromatography (LC) tandem mass spectrometry (MS/MS) system. The idea being that the target protein (or "bait") is pulled-down along with other proteins (or "prey") with which it interacts. These interacting proteins can then subsequently be identified and quantified. Modern LC-MS/MS systems (e.g., ThermoFisher Q Exactive) are capable of quantifying thousands of proteins in a single run, and the extraordinary sensitivity and accuracy of these instruments make them very effective at detecting constituents of PPIs. However, since the pull-down process is reliant on antibodies, poor epitope specificity can lead to the capture and identification of false positives that obfuscate true interactions with the target protein [3]. One solution to this problem is to use an affinity tag whereby the tag is fused to a gene of interest and endogenously expressed. The protein product bearing the tag can then interact natively with other proteins prior to lysis, and extraction of the tagged protein is performed with an antibody with very strong specificity to the tag. False positives can further be reduced by performing a two-step tagging process, referred to as tandem affinity purification (TAP) [4]. After the pull-down step is complete, the sample is processed and subjected to LC-MS/MS.

Raw data obtained from LC-MS/MS is first converted from proprietary vendor format into an open format such as mzML using software such as ProteoWizard [5, 6]. Software such as SearchGUI is then used to match MS/MS spectra to peptide sequences using popular open-source sequence database searching algorithms [7]. The peptide spectrum matches (PSMs) outputted by SearchGUI acts as input to the software PeptideShaker [8], which assigns confidence to PSMs, estimates false-discovery rate (FDR), infers protein identities from peptide data, and provides results for downstream viewing and interpretation.

With identified proteins in-hand, further computational analysis, such as with Significance Analysis of INTeractome (SAINT), is next needed to assign a confidence or probability of the likelihood of a true PPI [9, 10]. These results then need to be summarized and visualized in an understandable manner that provide biological insight. Multiple steps of data processing and computational analysis require knowledge of Unix shell scripting or programming in order to make these tools work together in a coherent pipeline, but for scientists without this type of background (e.g., a bench biologist), properly utilizing these tools in concert can be difficult.

One solution to this challenge is Galaxy, a web-based platform for reproducible and collaborative biomolecular analyses [11]. Use of Galaxy requires no prior programming experience, yet it allows users to string together a multitude of computational tools into pipelines or "workflows." User data, workflows, and history of any analysis are available in a persistent point-and-click interface, and these items are easily shared with other users or exported for offline analysis. A popular implementation of Galaxy that supports a variety of proteomic and multi-omic analyses comes from the Galaxy for proteomics (Galaxy-P) project (http://galaxyp.org) [12–14]. Galaxy-P supports SearchGUI/PeptideShaker workflows for sequence database searching and Automated Processing of SAINT Templated Layouts (APOSTL) workflows for downstream AP-MS data analysis and visualization [15]. Here we describe use of Galaxy-P and APOSTL for reproducible AP-MS data analysis in the context of the epidermal growth factor receptor (EGFR) interactome in non-small cell lung cancer (NSCLC).

2 Materials

For end-users wishing to follow the workflow contained in this chapter, all that is needed is a desktop computer (Windows, Mac, or Linux) with access to the Internet and a modern web browser (e.g., Chrome, Firefox, Microsoft Explorer). If needed, a copy of the Galaxy workflow presented here can be downloaded from the Galaxy Tool Shed (https://toolshed.g2.bx.psu.edu/view/galaxyp/apostl/0b6e9cc279fc), GitHub (https://github.com/galaxyproteomics/tools-galaxyp/tree/master/workflows/apostl), or at the provided Galaxy training instance (https://z.umn.edu/ap-ms).

The following software list is for advanced users (e.g., experienced Galaxy user, system administrator, bioinformatician), wishing to create custom workflows or install Galaxy and Galaxy-based AP-MS tools detailed in this chapter. All Galaxy tools detailed here are available from the publicly available Galaxy Tool Shed (https://toolshed.g2.bx.psu.edu/).

1. Unix-based operating system.
2. Galaxy-based tools (https://github.com/galaxyproject).
   (a) ProteoWizard tools (Tool Shed).
   (b) Protein Database Downloader (Tool Shed).
   (c) FASTA merge and replace (Tool Shed).
   (d) SearchGUI/PeptideShaker tools (Tool Shed).
   (e) APOSTL tools (Tool Shed).
3. ProteoWizard (http://proteowizard.sourceforge.net/).
4. SearchGUI (http://compomics.github.io/projects/searchgui.html).

5. PeptideShaker (http://compomics.github.io/projects/peptide-shaker.html).
6. SAINT (http://saint-apms.sourceforge.net/Main.html).
7. Python (version 2.7; https://www.python.org/).
8. R (https://cran.r-project.org/).
9. Shiny Server (https://www.rstudio.com/products/shiny/shiny-server/).

3 Methods

3.1 Identification of Affinity-Enriched, Interacting Proteins

The first module in the analysis pipeline involves the identification of proteins obtained from tandem mass spectra acquired from tryptic digests of proteins isolated by affinity-enrichment, followed by processing to characterize interacting proteins (Fig. 1).

1. Access the remote Galaxy instance at https://z.umn.edu/ap-ms. This instance provides testing data and software needed for learning this data analysis pipeline. *See* **Note 1** further describes the Galaxy interface.
2. To use this public instance, each user must register and create a login username and password. To register, click on the User tab, and then click on "Register." Create a username (must be formatted as a valid email address), password, and public name. Once registered, click on the User tab and click on "Login" with your user credentials.
3. The first step in the analysis is to import the required input datafiles. Once imported, these datafiles will become part of a History (*see* **Note 2** for more about Histories in Galaxy). Go to the Shared Data tab, and click on "Data Libraries." In the list of shared data, click on "APOSTL input." Select all files in this folder, and click on "Import to History." This folder con-

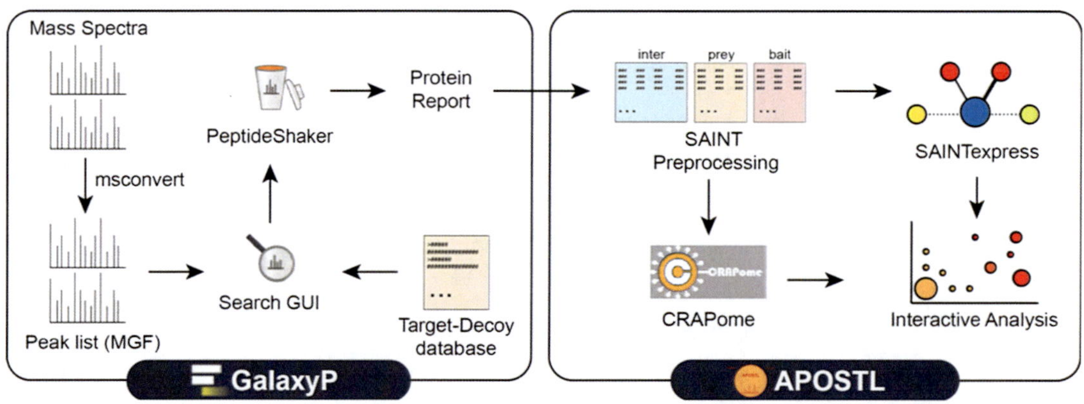

Fig. 1 Overview of the data analysis pipeline

tains four files in the Mascot Generic Format (.MGF), which are peak lists derived from MS/MS data files. It also contains a *bait* text file that contains information on the design of the AP experiment, which will be used for processing the results to characterize the interacting proteins. *See* **Note 3** for more on the nature of this file and the design of the data analysis.

To return to the main viewing pane, click on "Analyze Data." Each of these input files will now show up in the History pane as a separate item. If desired, the user can click on the History titled "Unnamed history" and enter a new title, making sure to hit "return" to save this new title. This will rename the active History.

4. With input data now ready in the active History, the next step is to import and run the workflow for analysis. Workflows contain all software and operating parameters necessary to carry out a specified data analysis (*see* **Note 4** for more about workflows in Galaxy).

   To select the workflow, click on "Shared Data" tab and select "Workflows." Then click on the dropdown arrow next to "APOSTL workflow" from the list. Select "Import." On the next screen click on "Start using this workflow." A page titled "Your Workflows" will now display. Click on the imported workflow and select "Run" from the dropdown menu.

5. A window now opens showing all steps that are part of the workflow. Select "Run" to run this workflow with the input files in the user's active History.

6. The Workflow will now open in the main viewing pane, showing the steps involved in the analysis (input/output data, software used). The Workflow automatically locates files of appropriate data type within the active History. Here, the ".MGF" files and "Bait.txt" file input files is recognized from the active History as the input data. Click on "Run workflow" to start the analysis.

7. After starting the workflow, new items will appear in the History pane. These are the outputs from the tools used in the workflow. *See* **Note 5** describes the software tools used in this workflow and outputs from each.

3.2 Visualization of Interacting Proteins

The second module in the analysis pipeline involves the visualization and interpretation of the data (Fig. 1). Using APOSTL, visualizations can either be static (Galaxy-based) or interactive (Shiny-based). Bubble graphs and dot plots are the only static visualizations supported, and users can use the tools available on Galaxy-P for this purpose (*see* **Note 5**). For interactive visualizations, user data must be first transferred to the Galaxy instance available at http://apostl.moffitt.org/.

1. For the sake of convenience, the *inter*, *bait*, *prey*, *list* (SAINTexpress output), and *CRAPome* files have already been transferred to http://apostl.moffitt.org/ and added to a shared History. To access the transferred data, simply navigate to http://apostl.moffitt.org/ and create an account if you haven't already done so. Then, click on "Shared Data" and "Data Libraries." In the new window click on "Galaxy-P and APOSTL." In the new window, highlight all files and then click the "Import to History" button at the top of the screen. The data will now appear in your History.
2. Under "Tools" on http://apostl.moffitt.org/, find and click "APOSTL Tools" to show the dropdown menu. Click "APOSTL Interactive Analysis," select the specified files from the History, and click "Execute." This tool generates an HTML file with a link to the interactive analysis in Shiny. Click on the "View data" (an eyeball icon) next to the "APOSTL Interactive analysis on data…" from the History and click on the provided link in the main window to open the interactive analysis. It is highly recommended to open the analysis in a new browser tab or window. The interactive analysis is powered by Shiny (https://shiny.rstudio.com/), a web application framework for the R statistical computing environment.
3. Various options for visualization and analysis exist within the interactive environment (*see* **Note 6**). For filtering, users can assign cutoffs or add proteins to an exclusion list directly from the sidebar. All plots are fully customizable with various themes. Plots update in real time as the user adjusts the filters or visual options. Raw data and visualizations can be downloaded at any point in various formats (PNG, JPG, PDF, TIFF, EPS, SVG) in 600 dpi resolution using the "Download Raw Data" or "Download Plot" buttons.
4. The data used in this chapter is from a study of the interactome alterations brought on by acquired resistance to the EGFR inhibitor erlotinib in NSCLC. We would expect to identify an interaction with EGFR and the adaptor protein GRB2 in an EGFR TAP experiment [16], and by using the interactive bubble graph we can confirm GRB2 is several times higher in EGFR TAP when compared to control (Fig. 2). For further biological interpretation of the results, users are referred to the original APOSTL manuscript, which details the rationale and experimental design [15].
5. Users are now equipped to perform their own AP-MS analysis using the Galaxy tools demonstrated in this chapter. The Galaxy-P and http://apostl.moffitt.org servers are provided free of charge for instruction and for processing very small datasets. Datasets containing more than a few baits will likely overwhelm these servers and may be canceled without notice.

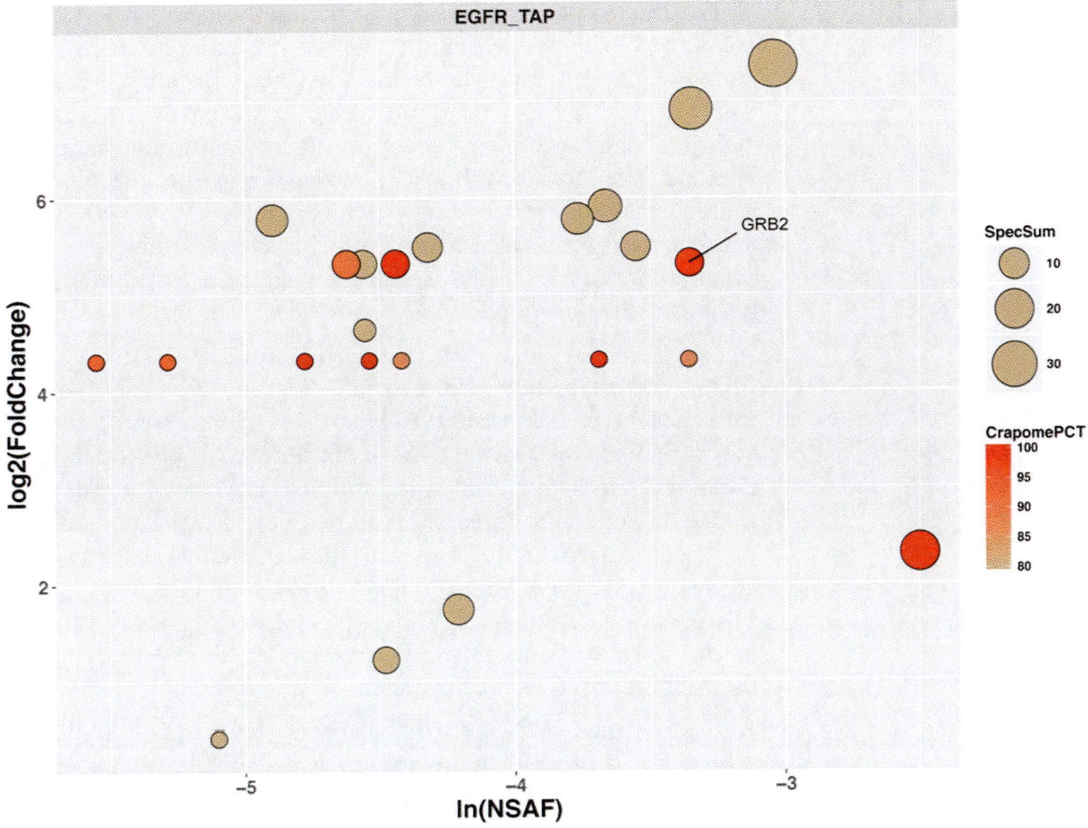

Fig. 2 Example of graphical representation of the results, which shows GRB2 is much more abundant in the EGFR TAP experiment. *NSAF* normalized spectral abundance factor, *CRAPomePCT* percent probability of specific interaction based on CRAPome database, *SpecSum* sum of spectral counts across all analyses

For help setting up your own Galaxy server with Galaxy, please see the Galaxy Project's administration page (https://galaxy-project.org/admin/) and the APOSTL GitHub page (https://github.com/bornea/APOSTL).

4 Notes

1. Galaxy uses a web-based user interface accessed by a URL. The URL can direct users to either a locally installed instance, or an instance running on a remote server. The Tool Pane (left side column) contains a list of available software tools in the Galaxy instance. The center portion of the interface is called the Main Viewing Pane, and is used to set operating parameters for software tools, edit and view workflows comprised of multiple tools, and also view results from data analyses. The right side column of the interface is the History Pane, which show the active History.

For the purposes of training in the methods described here, we have developed a training Galaxy instance at https://z.umn.edu/ap-ms. In order to use any Galaxy instance, it is necessary to register as a user and create login/password credentials. NOTE: accounts on the https://z.umn.edu/ap-ms training site are temporary. Accounts may be periodically erased, making it necessary to reregister to access training workflows and materials. Access to this site is anonymous and email addresses are only used for registration to the instance and not other purposes.

2. A record of any analysis in Galaxy is stored as a History (*see* https://galaxyproject.org/tutorials/histories/ for more information on Histories). A History is comprised of all the software used in the analysis, complete with the operating parameters used for each tool, and also all input, output and metadata generated. Histories can be of any length, containing just a few analysis steps and data files, or many. Histories are always saved, even when a user chooses to generate a new History for a data analysis. The active History is always shown in the History Pane. Any History can be shared with other users of the same Galaxy instance.

3. APOSTL uses SAINT (SAINTexpress) to assign confidence scores to protein-protein interactions based on provided AP-MS data. SAINT and therefore APOSTL requires three plain text files as input for assigning confidence scores to the AP-MS data: an *inter* file, a *bait* file, and a *prey* file. The *inter* file should have four columns of text (separated by tabs): IP names, bait protein names, prey protein names, and the quantitative AP-MS measurements (spectral counts, number of unique peptides, or intensities). The bait file should have three columns of text (separated by tabs): IP names (same as *inter* file), bait names (same as *inter* file), and the experimental design labels (target vs. control labeled as T or C). The *prey* file should contain three columns: prey protein names (same as *inter* file), protein length (number of amino acids), and prey gene names.

   The *bait* file is essentially the experimental design that is used to run SAINT, and in the case of the workflow presented here, it is used to label the input datasets in the *inter* file during SAINT preprocessing. With other types of input data (Scaffold, MaxQuant), APOSTL uses the baits to find the correct columns to generate the *inter* file since all of the samples are in the same file. However, with PeptideShaker outputs each sample is a different file, so the user is required to input the files in the same order that is specified in the *bait* file. This approach makes it such that the first input file is the first line in the *bait* file and so on.

SAINT has many requirements for properly formatting the three input files. APOSTL will aid in the creation of these input files, but issues can still arise with user-provided labels and data. Most issues that a user encounters can likely be corrected by rigorously checking that the provided input files meet the file formatting requirements. For a more complete description of SAINT input files, users are referred to a SAINT protocol and the SAINT manual found through the SAINT project's webpage (http://saint-apms.sourceforge.net/) [17]. In general, avoid hyphens and whitespace characters such as spaces in the naming convention.

4. Workflows include all the software tools used in an analysis, but without the input and output data (*see* https://galaxyproject.org/learn/advanced-workflow/ for more information on workflows). As such, workflows are different than Histories. Importantly, workflows contain the required parameter settings necessary for running the analysis on specific input data. Like Histories, workflows can be shared with other users of Galaxy. The ability to save validated workflows, and share these, makes for an efficient way to carry out a complex data analysis, avoiding the need for step-by-step optimization of parameters required for optimal results.
5. The software tools and data outputs used in this workflow are described below:
   (a) *Protein Database Downloader (PDD)*. This software tool imports protein sequence database (in the FASTA format), which are used for matching MS/MS spectra to peptide sequences. In this workflow, PDD is used twice: First to download the most recent version of the human protein database from the UniProt/SwissProt database; second, PDD is used to download the cRAP (contaminants) database, which contains common contaminant proteins detected in tandem mass spectrometry-based experiments. NOTE: For the purposes of the training workflows we have made available on https://z.umn.edu/ap-ms, we are using a static version of the UniProt protein database (UniProt release 2017_04, downloaded on April 12, 2017). Using updated versions of this database may result in slightly different results from this workflow.
   (b) *FASTA Merge Files and Filter Sequences*. This Galaxy tool concatenates separate FASTA protein sequence database files, removes any redundant entries, and outputs a single sequence database. Here, this single FASTA database file is the combination of the human proteins and the common contaminants.

(c) *SearchGUI.* This tool performs protein identification by matching MS/MS peak lists (Mascot generic files, .MGF) with sequences contained in the FASTA file. SearchGUI bundles popular open source sequence database searching programs. For the workflow used here, relevant parameters are:

- Sequence database search programs used: X!Tandem, MS-GF+ and OMSSA Parameters for each of these are set and saved within the stored workflow.
- Protein digestion parameters: Trypsin, with two maximum missed cleavages.
- The precursor ion tolerance is 10 ppm, with fragment tolerance of 0.5 Da.
- Minimum/maximum charge of ions: 2/6.
- Fragment ions searched: *y* and *b*.
- Fixed protein modification: carbamidomethylation of C.

(d) *PeptideShaker.* PeptideShaker processes the output file from SearchGUI. It infers proteins from matched peptide sequences and applies statistical metrics to assign confidence to identified peptides and proteins. Within this workflow, the "Advanced" options are selected, with relevant parameters as follows:

- The maximum FDR value (%) at protein level is 5.0, peptide level is 5.0, and PSM level is 5.0.
- Minimum and maximum peptide length are 6 and 75, respectively.
- Maximum precursor error is 10.0 ppm.
- Outputs selected: PSM report (tabular), Protein report (Tabular), and Certificate of Analysis (text).

(e) For the workflow used here, each single MGF file is separately analyzed by SearchGUI/PeptideShaker, with separate outputs generated for each and then used as input for subsequent characterization of interacting proteins.

(f) *SAINT Preprocessing.* This tool will read in PeptideShaker protein report files and process them to generate an *inter* file, a *bait* file, and a *prey* file required for SAINT (SAINTexpress) analysis. This tool can also accept Scaffold samples reports exported as plain text files and MaxQuant peptides.txt files. Users have the option to provide their own *prey* and *bait* files or have them generated with the aid of APOSTL.

(g) *Query CRAPome.* This tool takes the *prey* file from the SAINT Preprocessing tool, compares against the CRAPome human contaminant database [18], and returns a file specifying the prevalence of each protein present in the CRAPome.

(h) *SAINTexpress.* This tool analyzes the *inter*, *bait*, and *prey* files output from the SAINT preprocessing tool using SAINTexpress. Unless you are an advanced user, it is suggested that you utilize the default settings (100 replicates; 1 virtual control). The output is a file containing individual bait:prey interactions on each line. The interaction is scored by comparing against the negative controls (values of which are in the ctrl_count column). Other characteristics included are (but not limited to) the SaintScore (probability of a true interaction), Fold Change (over controls), and the Bayesian False Discovery Rate.

(i) *APOSTL Static Bubblegraph Generator.* This tool is used to create noninteractive bubble graphs for visualizing the SAINT results. A typical visualization includes plotting ln(NSAF), the logarithm of the normalized spectral affinity factor, on the x-axis and $\log_2$ fold change on the y-axis. *See* **Note 6** for suggestions and interpretation of these visualizations.

(j) *ProHits DotPlot Generator.* This tool runs the biclustering version of the DotPlot tool for visualizing multiple baits simultaneously (found at http://prohitstools.mshri.on.ca/) [19]. An interactive version of this tool can be found at http://prohits-viz.lunenfeld.ca/.

6. Many options exist for each interactive visualization in APOSTL. General descriptions and tips are described below:

(a) Correlations: This tool compares two user specified replicates and displays a scatterplot of the data with the R-squared value. This box's position can be adjusted using the sliding bar with the left adjusting the x and right adjusting the y positions.

(b) Boxplots: This tool compares the raw spectral counting or intensity values of all individual bait groups in the dataset as a boxplot.

(c) Density plots: This tool analyzes the distribution of the ln(NSAF), $\log_2$ fold change, NSAFscore, SaintScore, SpecSum, or the LogOddsScore across all individual bait groups. Baits can be added and removed from visualization.

(d) Bubble graphs: Bubble graphs are the most popular visualization in APOSTL. This tool splits graphs into individual test bait groups and visualizes these bait groups simultaneously as a modified scatterplot. Users can select ln(NSAF), Log_2 Fold Change, NSAFscore, SaintScore, SpecSum, or the LogOddsScore for the *x*- or *y*-axis as well as for the size of individual bubbles. A typical visualization includes plotting the ln(NSAF) on the *x*-axis and log_2 fold change (relative to the control) on the *y*-axis. The color of the bubbles can be "fixed" so that all bubbles have the same color, or scaled based on individual prey's prevalence in the CRAPome (default). The values used to scale bubble radius can be adjusted. Users have the option to label either all bubbles present in the graph or all the bubbles passing an 80% CRAPomePCT cutoff. Labels are force directed meaning that the tool will pick label placement to avoid overlap.

Higher abundances should indicate stronger (and more confident) interactions, so ln(NSAF) allows for the prioritization of proteins that have a higher abundance in their enrichment. Indeed, SaintScore and fold change are both susceptible to false positives from hits being identified in the enrichments at a low abundance and not in the controls. This can happen often with large proteins or sticky proteins; this can happen because common controls aren't perfect.

(e) PPI network: all interactions passing the filtering criteria specified in the sidebar are visualized as connections between the bait and individual preys. By default, physics are applied to the network, which allows for the network to self-arrange. This network can be downloaded as a PNG (using the button within the graph pane), as a simple interaction file (SIF), or as a GEPHI JSON file. For more control over the network visualization, it is recommended to import the SIF or GEPHI JSON file into Cytoscape or GEPHI.

(f) Bait2Bait (2D-histogram): This tool compares individual bait groups directly to identify bait groups that are most similar. Shared interactions are scored for each bait-bait comparison and the resulting matrix is clustered to arrange similar baits together.

(g) KEGG Pathways and Gene Ontology: These tools query KEGG and Gene Ontologies using the ClusterProfiler R package and visualizes the output as a bar graph [20]. Since the calculations for these tools can take up to 30 s, an "Analyze" button is included where the user specifies the options and then clicks "Analyze" to begin the calculations.

Acknowledgments

The authors acknowledge support from NIH grant U24CA199347 and NSF grant 1458524 to the Galaxy-P team members (P.K., S.M., J.J., P.J., T.G.), the Moffitt Lung Cancer Center of Excellence (P.S.), and the NIH/NCI F99/K00 Predoctoral to Postdoctoral Transition Award F99 CA212456 (B.K.). This work has been supported in part by the Biostatistics and Bioinformatics Shared Resource at the H. Lee Moffitt Cancer Center & Research Institute, an NCI designated Comprehensive Cancer Center (P30-CA076292).

References

1. De Las Rivas J, Fontanillo C (2010) Protein-protein interactions essentials: key concepts to building and analyzing interactome networks. PLoS Comput Biol 6:e1000807
2. Scott DE, Bayly AR, Abell C et al (2016) Small molecules, big targets: drug discovery faces the protein-protein interaction challenge. Nat Rev Drug Discov 15:533–550
3. LaCava J, Molloy KR, Taylor MS et al (2015) Affinity proteomics to study endogenous protein complexes: pointers, pitfalls, preferences and perspectives. BioTechniques 58:103–119
4. Gregan J, Riedel CG, Petronczki M et al (2007) Tandem affinity purification of functional TAP-tagged proteins from human cells. Nat Protoc 2:1145–1151
5. Jones AR, Eisenacher M, Mayer G et al (2012) The mzIdentML data standard for mass spectrometry-based proteomics results. Mol Cell Proteomics 11:M111.014381
6. Kessner D, Chambers M, Burke R et al (2008) ProteoWizard: open source software for rapid proteomics tools development. Bioinformatics 24:2534–2536
7. Vaudel M, Barsnes H, Berven FS et al (2011) SearchGUI: an open-source graphical user interface for simultaneous OMSSA and X!Tandem searches. Proteomics 11:996–999
8. Vaudel M, Burkhart JM, Zahedi RP et al (2015) PeptideShaker enables reanalysis of MS-derived proteomics data sets. Nat Biotechnol 33:22–24
9. Choi H, Larsen B, Lin ZY et al (2011) SAINT: probabilistic scoring of affinity purification-mass spectrometry data. Nat Methods 8:70–73
10. Teo G, Liu G, Zhang J et al (2014) SAINTexpress: improvements and additional features in Significance Analysis of INTeractome software. J Proteome 100:37–43
11. Afgan E, Baker D, van den Beek M et al (2016) The Galaxy platform for accessible, reproducible and collaborative biomedical analyses: 2016 update. Nucleic Acids Res 44:W3–W10
12. Boekel J, Chilton JM, Cooke IR et al (2015) Multi-omic data analysis using Galaxy. Nat Biotechnol 33:137–139
13. Jagtap PD, Johnson JE, Onsongo G (2014) Flexible and accessible workflows for improved proteogenomic analysis using the Galaxy framework. J Proteome Res 13:5898–5908
14. Sheynkman GM, Johnson JE, Jagtap PD et al (2014) Using Galaxy-P to leverage RNA-Seq for the discovery of novel protein variations. BMC Genomics 15:703
15. Kuenzi BM, Borne AL, Li J et al (2016) APOSTL: an interactive Galaxy pipeline for reproducible analysis of affinity proteomics data. J Proteome Res 15:4747–4754
16. Lowenstein EJ, Daly RJ, Batzer AG et al (1992) The SH2 and SH3 domain-containing protein GRB2 links receptor tyrosine kinases to ras signaling. Cell 70:431–442
17. Choi H, Liu G, Mellacheruvu D et al (2012) Analyzing protein-protein interactions from affinity purification-mass spectrometry data with SAINT. Curr Protoc Bioinformatics Chapter 8:Unit8.15
18. Mellacheruvu D, Wright Z, Couzens AL et al (2013) The CRAPome: a contaminant repository for affinity purification-mass spectrometry data. Nat Methods 10:730–736
19. Knight JD, Liu G, Zhang JP et al (2015) A web-tool for visualizing quantitative protein-protein interaction data. Proteomics 15:1432–1436
20. Yu G, Wang LG, Han Y, He QY (2012) clusterProfiler: an R package for comparing biological themes among gene clusters. OMICS 16:284–287

Index

A

B

C

D

E

F

G

H

I

J

M

N

P

Caroline A. Evans et al. (eds.), *Mass Spectrometry of Proteins: Methods and Protocols*, Methods in Molecular Biology, vol. 1977, https://doi.org/10.1007/978-1-4939-9232-4,

Zeitfracht Medien GmbH
Ferdinand-Jühlke-Straße 7
99095 Erfurt, Deutschland
produktsicherheit@kolibri360.de